HERE AND THERE
THEN AND NOW

By

PAUL HILL

Cover art: watercolor by Paul Hill.

Drawings and paintings by Paul Hill. The paintings are in color but, except for the cover, they are printed in black and white.

Some images courtesy Pixabay.com.
All rights reserved by the author.
Copyright © 2022

Publisher - William Serle
billserle@aol.com
billserle.com

Dedication

Many thanks go to family and friends and especially to Mary Jane Law, to my daughters Margaret Burrows and Georgia Leacox, to my Writing Workshop friends at Martin Andersen Senior Center, and to Bill Serle, my publisher. They helped and inspired me to create this book.

FOREWORD

Here And There, Then And Now by Paul Hill, includes vignettes from his interesting life. It is well-flavored with splashes of practical and exotic knowledge on a wide variety of subjects. It was written with humor and kindness. May it long entertain you.

Please, Dear Reader, delight in his words and even feel free to argue when you disagree. He'll listen to your point of view with respect and, even if he disagrees, he will always be agreeable.

I am proud to know and be a friend to Paul Hill. He is a big man who accepts the blows life delivers with grace and strength.

I wish you 'Fair winds and following seas,' Paul.

Bill

William T. Serle, Jr.
Author and publisher

A SHORT INTRODUCTION

I have been a storyteller since I was a schoolboy, and now, in my twilight years, I've been encouraged to write them down.

It is meant as entertainment.

Paul Hill

TABLE OF CONTENTS

Boats 1
Learning to Drive 7
The Flying Business 11
Sauerkraut 17
The slop chest 21
Trainman 25
Short Line Railroads 27
Miscibility 29
The Victory Ship 31
Blue Smoke and Mirrors 35
Shrimp Boats Are Acoming 41
Not A U-2 45
Feeble Excuses 47
Patents and Copyrights 49
I Wanted Wings 53
A Very Special agent 59
Prisoner Chase 61
The Dance Company 65
Patty and I 69
Faking Business 73
A Very Short History 77
Hard Coal and Soft Coal 79
Home Sweet Home 83
Ramblings 87
Crosley Autos 91

Albany the Capital	95
Newscasters	97
A Lot of Bull	101
Write and Wrong	105
The Masked Man	107
Warping Tugboats	109
Self-Employed – An American Tale	113
Charlie	117
Cargo Handling	121
A Young Farmhand	123
A Small Railroad Town	127
Tropical Storms	129
What Time is Noon	133
Heroes I Have Known	135
Plimsoll Marks	141
A Tale of Monkey Fists	143
Birds of a Feather	145
Fire Alarm Systems	151
Iceboating	153
International Code of Signals	155
The Fourth Estate	157
Hog Islanders	161
Reverse Mortgages (Good or Bad?)	163
The Draftsman	165
Aircraft I Have Known	169
To Sea or Not to Sea	173
A Change of Diet	175
My One Lung Skiff	177
Parade Dress	181

Figureheads and More 183
Knowing Your Ropes 187
My U-2 Story 189
A Big Secret – How We Won
　　World War II 195
Hole in the Blockhouse Wall 201
Watervliet Arsenal 205
Hardship Duty in New York 211
Fire Hydrants and More 215
Tobacco Then and Now 219
Changing From Chain 223
My Last Sailing Race 227
Changes in the Workforce 233
Spring is Here 235
Radio Versus Television 239
Police Radios 243
Tatting and Such 245
A Joke 247
Clouds 249
Candy Cane Lodge 253
The Chicken Business 255
The Time of my Life 259
The Write Instruments 263
A Reunion (Or Farewell Gathering) 267
Jeep 271
Health and Happiness 273
A Ride in a Glider Aircraft 277
The Rise and Fall of USA Railroads 281

World War II
    (The Loon Weapon System) 283
Powder Monkey 285
The Caterpillar Pin 287
Trimming Grain 291
A Borrowed Wheelbarrow 295
Art vs Illustration 301
Aunt Kitt the Piano Player 299
Build a better Workbench 301
Half Tracks 305
Steam Versus Gasoline Engines 307
No Working Title 309
True North 313
Small Boats Now and Then 315
Balance of Power 319
North Country Crafts 323
The Day I Met Tarzan 327
Weather of Not 329
My Different History of
    Missile Development 331
Unsung Heroes 335
The Wright Brothers 339
Feathering Propellers and
    Paddle wheels 343
About the Author 349

"Skipjacks" – Watercolor by Paul Hill.

BOATS

I grew up around boats and ships, where the Mohawk and Hudson rivers meet with canals heading north and west, amid numerous lakes and streams. Watercraft were everywhere. From the smallest kayak to the ocean freighters and tankers, one could not help being impressed by the variety. The Hudson was the busiest river on the East Coast. Commercial shipping and private craft were in abundance.

I specifically delighted in the Hudson River Day Line steamboats. Each year we have a school excursion on one of the last four of the scheduled steamers.

The Hendrick Hudson was a screw boat. The Robert Fulton, the Dewitt Clinton, and the Alexander Hamilton were paddlewheel driven. These canoe-shaped wrought iron hulls sliced through the waters from Albany to New York, making stops along the way.

Licensed to carry 4,000 passengers, they had exquisite furnishings with hardwood furnishings and tapestried walls. They were floating palaces.

My family had a variety of small craft over the years. At the bottom of our street was located one of several boatyards. Private boats and small commercial craft were built and serviced here.

My first engagements were caulking and painting wooden rowboats. My older brother Bud (Anthony J. Hill, Jr.) bought a kit, and we built a fabric-covered kayak.

Itching to make my own, I wrote to the Douglas fir plywood association and obtained a free set of plans for an eight-foot pram. Made from two sheets of plywood and wood frames, it took two weeks to put in the water.

I built and sold a half dozen copies during my high school years. A few were still in use many years later. Bud and I constructed a comet sailboat, which he kept for years.

After high school, I went to sea for two years and then was drafted into the military.

It was inevitable that another boat came around. Newly married and living in Florida at this stage, I began working in the space programs. While staying around marinas and doing odd jobs, a sailboat found me.

A couple living on a large ketch had a small sailboat, which was damaged. They were leaving the area for a new assignment. So, they sold it for a pittance. The boat was built in the Bahamas. An excellent wooden craft.

The 14-footer had its side stove-in by an auto backing into it while on shore. I took it as a challenge to replace several steam-bent frames and replace clinker-style planks using copper nails and rawls. Shirley, my wife, and I enjoyed the hours spent on the Indian River. She was of Norwegian descent.

We sold the sailboat to two brothers. They had joined the Coast Guard rather than the Navy, hoping to get time in small ships. As intelligent young men,

they were sent to work on electronics at shore stations and never touched a boat. So, they bought their own craft.

From 1958 to 1968, I was active in the Coast Guard Auxiliary. Rising through the ranks, I was selected Flotilla Commander. Being immersed in things nautical, we have many friends and memories. Building in fiberglass was taking over small boats.

A Mr. Palmer built Rebel sailboats and sold fabric, mat, and resin to home builders. Fate intervened. The sea scouts had a mold for a 16-foot cargo canoe. It was to be shipped to Tampa the next day.

One long night, two buddies and I molded three canoes.

Each took home a blank hull to be completed with rails, breast-hooks and thwarts. My canoe was painted and in the water by the weekend.

The next boat was much larger and an ambitious speculation. An older U.S.C.G.A. member passed, and his wife wanted me to take over his boat. It was a twin-screw 34-foot sports fishing boat, being repaired.

The vessel was in Vero Beach undergoing extensive repairs. I commuted weekends and made ready to bring the boat to Port Canaveral. Finally, we would leave the Vero and head north. I was settling accounts at the fueling dock when a large boat came to fuel.

I recognized the aluminum-hulled yacht as Bill Evinrude's. A man approached and asked if we could move so they could fuel

I immediately recognized the suntanned man in sports clothes as Billy Graham, the evangelist. He sure was handsome.

"May I shake your hand?" I asked.

Without further to do, we came up the intercostal waterway. Weeks passed.

Working at the Kennedy Space Center, volunteering at the Coast Guard station, and fixing the boat made for many long days and nights.

Finally, with visions of Crunch and Des capturing our imagination, we went out into the Gulf Stream. Hoping to do well by taking parties fishing offshore proved unsuccessful.

You see, I could operate and maintain the boat but, I was not a fisherman. Parties don't come back if you don't catch fish. We sold the boat and went on to other ones.

Over the years, I built an inboard Aikten design, several clamming skiffs and overhauled a 22 Catalina sloop while helping others.

Now elderly and retired, my love of things to do with boats consists of marine artwork, writing, and building model boats.

These models are purposely crude. They are meant to represent handmade, using few tools, as a retired builder might do.

In closing, I remember the Water Rat tells Toad,

"Believe me, my young friend, there is nothing – absolutely nothing – half so much worth doing as simply messing about in boats."

LEARNING TO DRIVE

When I tell young people that I learned to drive a Ford Model AA truck, they try to correct me. Just as everyone calls every light plane a 'Piper Cub,' they also know that all Fords built in 1931 are 'model A's.' For the record, a Ford Model AA truck was a one-ton truck, built with a heavy frame, and dual rear wheels. While the engine and bodywork were common to the popular cars and pickup trucks, there were differences.

The rear-end (differential) had low gear ratios, and the transmission was a five-speed, allowing the 28-horsepower motor to pull heavy loads. I mention these peculiarities, as they limited performance and required special driving skills.

Now, for the driving. We prepare to drive by making sure the parking brake is on, and the gear shift is in neutral. Then, set the manual spark advance and the throttle, located below the steering wheel. Reaching over to the combination gas shutoff, choke, and mixture control turn the ignition key and start the engine.

This requires practice, especially in north country during cold weather. It must be done in a gingerly fashion and with patience. Common practice requires you just sit and wait several minutes for the huge amount of oil in the crankcase to warm up before moving. The shifting pattern, for this five-speed, necessitated a thumb latch to be raised, allowing the reverse position to be selected by knocking any passenger's knee.

I should mention at this time, that my father was a self-employed welder. That is, he had mounted a Chrysler-powered, Hobart welder, with acetylene outfit and tools, permanently on the rear chassis. Thus, this one-ton truck every day carried two tons. If one is to drive this rig, you must adopt a defensive attitude. Also, with no power steering, you develop upper body strength. Consider the foot-powered mechanical brakes, and you'll learn to be very watchful and plan your every move.

Now, depress clutch, shift into first gear, and release handbrake. Turn signals were after-market accessories; so, we use lots of hand signals. The business of getting underway and mixing in traffic could be adventurous and nerve wracking.

Our cab was fitted with a Saint Christopher's medal. You see, flat-out foot-down, with a tailwind) this over-loaded four-cylinder truck couldn't go forty-five miles per hour. The general rule was stay to the right and don't even think of passing. Do not tailgate (a definite no-no with mechanical brakes)! Always leaving several car lengths ahead of you causes a constant stream of vehicles passing you and darting in front of you.

Let me mention some special features, like the cabin heater. It consisted of a tube fitted around the exhaust manifold and producing only exhaust fumes. The windshield wipers were vacuum powered. This requires special consideration. When going uphill, you may be required to shift to a lower gear. This entails 'double-clutching.' By relaxing the powertrain, manifold pressure rises, and the wipers stay to the upper right and refuse to move (and rain/snow covers your field of vision).

The primitive electrical system, providing power to the headlamps, with their soft yellow glow illuminated about ten feet in front of the vehicle. This could make night-driving an adventure.

In a 1931 vehicle, there were no creature comforts as air conditioning or adjustable seats, etc. There was a slight consideration: the windshield swung forward, allowing the forty-mph breeze in. But, we were strictly forbidden to open it. You see, this 16-year-old truck, with rotted seals, would leak all around the windshield; so, my father had caulked it shut and removed the hardware. Did I mention the mixture control and the spark lever, which required attention? How about pulling to the side of the road and putting on a set of chains, during snow and sleet – at night – on unplowed roads.

Well, as you can see, our worries weren't speeding tickets; but rather, obstructing traffic. It shouldn't surprise you to hear that a round trip to the county seat (eight miles away) may take half a day. But that Ford was all we had. There were no other trucks available for civilian use during WWII.

After the war, we finally got a much improved late-model truck, which was a pleasure to drive. My brothers and I thought we were in heaven. In recent times, there are some people who fancy to own an old Ford truck for a hobby. They never had to earning a living with one.

THE FLYING BUSINESS

Being a schoolboy during World War II, and a C.A.P. Cadet through high school the romance and glamour of being a pilot was all consuming.

My buddies and I would sell our souls to get to fly those loud, metal machines up into the sky. The vastness and the freedom were intoxicating.

Well, as the years went by, and reality set in, I matured, and life's demands showed up. The aviation business became tarnished and less inviting.

During my experience in the Air Forces 'Tiger program, I satisfied the requirements for an ATR rating. That is 'airline transport rating' with IFR, 'instrument flying rules,' training.

Two incidents happened while in training. First, a colonel came to our base flying an F-86. All the students surrounded this latest jet fighter. Given a chance to sit in a frontline jet was a thrill.

But nature stepped in I measured 48-inches tall from seat to top of the head. Wearing a helmet. Seated on a parachute, I would not be able to close the canopy. "You won't be a fighter pilot," I was told.

The second scenario involved the thrill of a familiarization flight in a twin-engine bomber.

The Douglas built A-26 was offered in several configurations. It is sometimes confused with A-20 or B-26. Rejected by our air services, the first 200 were bought by France. They proved very effective.

When France fell, the Germans took over the planes and used them against the allies at Casablanca. Later they were adapted by American forces.

I describe radial 'roundie' engines elsewhere. Visualize the well-known B17 heavy bomber, with four engines on its wings. Look at the controls and notice four of every engine instrument, in four quadrants, with three levers for each engine. (Lots of confusion.)

The aircraft being described, the A-26, had twin row radials – that is 18 cylinders, developing twice the power. This smaller bomber was much faster than its big brothers.

Also, when I mention it being operated by the single pilot, people may not believe me. When I further remark the RAF's famous Lancaster bomber was flown by a single pilot, usually a Sergeant, they may go back to their books. However, the few A-26 intruders used to train pilots to handle them alone, were equipped with dual controls.

In combat roles, most A-26s were attack bombers, in support of ground forces. Flying at low altitudes, strafing, bombing, and using rockets. Some were used night intruders, painted black flown as single-aircraft reconnaissance missions.

This powerful, small bomber was flown as a fighter, and could be horsed about the sky great in evasive maneuvers. They were kept in service long after most WWII planes were gone. The last actual combat for these in hot aircraft, was the ill-fated Bay of Pigs Mission.

Military aircraft of World War II were rugged and expected to be stressed. They were all noisy, inside, and out.

When inside those ships, headsets were worn like earmuffs and throat microphones were necessary. There were no creature comforts, no upholstery, insulation, heat, or air. The framing was left exposed, in the interest of cost, weight considerations and repairs.

High altitude crews wore arctic clothing or electric suits, if available.

Military aircraft seldom transferred to civilian roles. Some aircrafts, that were civil aircraft used by the forces, did return.

The most famous 'Gooney Bird Douglas DC-3 was made in thousands as the C-47 transport.

Airliners were stripped of their interiors, fitted with bench seats and a wide cargo door. Everyone used them, even our enemies. A few dozen are still flying. Over 13,000 were made.

I was a pilot of a C-47, used in charter work. As an unpressurized aircraft, it was not to be flown above 12,000 feet. The mountains rose to 6,999 feet. We were up in the weather, all the time.

For an economical cruise we flew at 165 mph. This ship had fuel capacity for ten hours. Also, it was steam heated.

The old fuel system had the gasoline from the wings, sent to the flight deck. Manual valves were then used to route fuel to the engines. As taildragger one couldn't forget to lock the tailwheel before takeoff.

Also, there were no control actuators 'power assist. Muscles were tired after long flights. Some people speak fondly of the C-47 and the C-3, and steam locomotives. They didn't have to live with them.

Another popular airplane, built by Beechcraft, was model D-18. With military designation C-45 it was configured to be a bombardier and a navigator trainer this ten-place twin engine is still seen with its twin tail.

The Lockheed 'Constellation,' designed by Howard Hughes, is considered to be the most beautiful transport ever built. It was not widely used by the military for one simple reason – it was too expensive. Travel posters of 'Connies' are collector's items.

In recent years, several major changes have radically made flying a totally different world.

The motivation for being a pilot is not the same. The gung-ho spirit of the Flying Tigers, and the John Wayne and Robert Stack films, are all gone to the Hangar in The Sky.

Large airliners carry hundreds of people at near sonic speeds. The formula for passenger seat miles per pilot is grown geometrically. Fewer pilots are needed.

'Rosie the Riveter' is retired. Airframes are chemically welded and glued together. Electronics manage fuel, navigation, and help with traffic control.

Crews have changed. The first to go was the radio officer. The next was the navigator. Then the flight engineer.

A joke in circulation, tells us that the crew of an airliner in the future will consist of a pilot and a dog. The pilot's main duty will be to take care of dog. The dog's duty will be to bite the pilot if he touches anything.

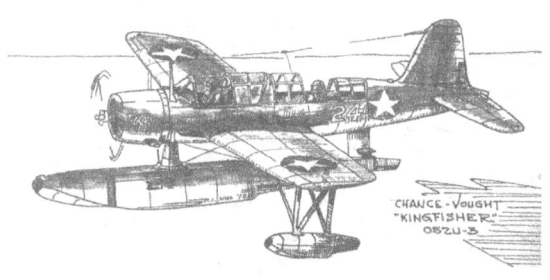

Dave and I were on USS North Carolina (BB-) and examined a Kingfisher on a catapult at Moorhead, North Carolina. This A/C was spot-welded.

SAUERKRAUT

The build-up to World War I was long coming. Common knowledge. The problems that existed in Europe only needed a spark to ignite a major conflict.

The first modern war, the American Civil War, caused the development of mechanized war materials. War marshals were itching to try out rapid-fire guns, steamships, lorries, and even poison gas.

People took the hint and left Europe by the groves.

George Bender grew up near Frankfurt where the zeppelins were built as weapons of terror.

He took a ship to New York, where the Dutch communes historically existed. He finally settled in Albany's 'Dutch Hollow. Being a smart fellow and ambitious he went into business for himself. One thing he did know was sauerkraut.

He bought several adjacent pieces of real estate as the years passed. The store he used for business had a flat above where he, his wife and children lived. The framed two-story buildings had common walls providing for low-rent flats.

By breaking through two basement walls, they provided a Kraut cellar three basements wide. Machinery for cutting cabbage and hundreds of hog's heads full of salted kraut, was weighted ay down by cobbles.

Upstate New York was a great farming area, steeped with a history of Rip Van Winkle, Sleepy Hollow and such. The rhythm of life abides with the seasons of the year

The earliest crops brought the best prices. As each year came under threat of winter, the hardiest crops had to be sold, or lost. George bought cabbages by the truckloads at his prices.

The cabbages were made into Kraut and stored beneath the rented flats. When ready, the product must be sold. George proved to be a good salesman. Wooden buckets were used by the many beer halls in the free lunches and the restaurants. He especially made deals with hospitals and jails.

In the days before frozen food, many food stuffs were smoked or pickled for use during the winter months. Over the holidays and specially during the Lenten season, the cellar was emptied and cleaned.

After Easter, the store was kept busy by selling companion products like pickled herring, cucumbers, pretzels and chips, and cheeses.

Oh, the cheeses! My brothers and I were often asked to make the first cut on the huge heads of cheese. Pulling the wire required great strength, making pieces that the women could handle. People offered 'big eye Swiss cheese. We found bubbles on one side only. Then we used a kitchen tool to cut balls out as our reward for helping out.

Spring was followed by summer and the business cycle ended.

An accountant would show whether George's efforts were profitable.

Doing the depression labor was cheap and bargains could be found. George bought a piece of riverfront adjacent to the Saratoga Battlefield.

Pruyn Lumber Company built a two-car garage without the large doors. The space was partitioned into two bedrooms – his and hers, living room and small kitchen. A screened porch spanned the front.

A well was sunk and a two-holed outhouse with star and crescent cut in the doors, was built. Lighting was by oil and mantel lamps.

The riverfront was improved and rowboats, with caulked seems were made.

Now passed middle age, George looked and acted like doughty burgher he was. With a large beer belly and a walrus mustache, his vest sported a gold pocket watch and chain. Among his collection of pipes, he preferred a meerschaum, and acted the Baron he was.

He loved people and was very kindly, especially to children. As 'Opah he ruled over his extended family through each summer.

A routine used by couples with children was to leave the heat of the city and come up to the 'summer place. After a weekend of picnic and fun, the parents returned to work and the young folks stayed, to be exchanged next weekend.

The upper Hudson River was a delightful place. Besides the canal traffic, yachts of all descriptions pass by, especially during the season at Saratoga springs, to dock at Schuylerville.

Let me explain, the Famous Racing place, Saratoga Springs, is six miles west of the river. The historic battle was along the river and old Saratoga was renamed in honor of General Schuyler, Alexander Hamilton's father-in-law.

George Bender died a few years after World War II. The sauerkraut business was regulated out of existence. His many heirs lost interest in the 'camp.

The next generation didn't like to vacation at the same place each year. The mobile generation had lots of options to visit.

My father took possession of the camp. With five grown sons, all mechanics, we made many improvements. Electric lighting, plumbing, radio, and telephone. Heating made the use of the place possible for hunting season and winter sports. As years went by, the area developed. The road was paved. Permanent homes were built. One by one the summer folks sold out

No one objected when my parents sold out. They bought a place on Merritt Island, Florida where they enjoyed their final years.

But I still think of George Bender and the good old days whenever I have a Reuben sandwich with sauerkraut.

THE SLOP CHEST

In the days of great sailing ships going on long voyages that could last years, – such as whaling ships, some of the crew did not return due to accidents, desertion, or other causes.

The missing crew members belongings were to be protected. If he had a family, say back in New England, his gear would be auctioned to the crew. The proceeds were sent to them. If that custom could not be carried out, his clothes and belongings would be kept in a chest and used for the welfare of others, as directed by the captain.

Captains still manage this practice, but in the modern fashion. The slop chest is no longer a chest like a footlocker, but a small room on a ship with a stout door and a heavy lock

In practice, the slop chest serves several functions. It fills the needs for personal items, like work clothes, foul weather gear, toiletries, razor blades, toothpaste, etc. Also, some ships can supply candy and cigarettes. Items are sold to the crew at no profit. The actual operation may be handled by the Stewards Department – and not the captain.

I should mention sea store cigarettes. Back in the 1950s, I personally bought cartons of cigarettes for $0.80 a carton. That's $0.08 a pack.

These packs provided by ship chandlers had no taxes to be collected. Instead of the most common blue paper tax stamped, each pack had a white paper strip, with a notation, 'For Consumption on the High Seas. Seamen were cautioned that there were penalties for landing and trafficking in cigarettes.

Today, the U.S. Navy has ship service stores contractors.

Another consideration regarding customs duties involves bringing goods in and out of different countries. It varies the world over. If one has goods that are not to be declared to customs, they are kept in the slop chest under lock. An official stamp is glued over the Keyhole. Upon leaving the territorial waters, of the country visited, the seal must be destroyed.

I shipped out during the 1950s. Many practices have changed. I wonder if the term slop chest is still used.

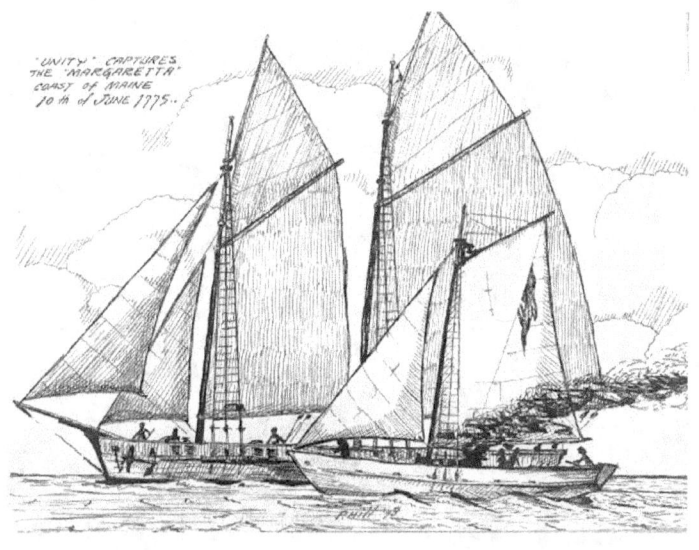

TRAINMAN

The oldest continuous railroad in the United States was the Delaware and Hudson railroad. It ran from the coal fields of Pennsylvania to the border of Canada and was divided into two divisions. From Honesdale to Albany was the Susquehanna, and from Albany North, the Champlain division. Headquarters was in the plaza building, at the foot of State Street, the hub of the capital of New York.

Myself and two buddies were interviewed by the trainmaster and hired as Trainmen on the Extra-Board. Let me explain.

The D&H was basically a coal road. It owned the hard coal mines and ran their locos on anthracite coal. The road also ran freight and passenger service, but some of their business was seasonal. Therefore, the Extra-Board

Permanent employees of the railroad were unionized and ran the regular operations. But for unscheduled and seasonal overloads, casual help was allowed. We were not expected to pay or join the union unless put on permanently.

Because all three of us grew up in around railroads, we had basic knowledge of railroading. We were closely supervised by the road foreman – at least for a while.

The duties of a Trainman are tiresome and dangerous. Hiring out on the extra board meant you had no regular runs and did not get paid by the hour. The pay was $21.10 per hundred miles.

It could take eight hours for a shift, maybe more. Depending on traffic and the crew, you could go home early if you completed your assignment earlier.

My experience was only on freight trains, only at night, and as a 'headed man;' that meant riding in the cab of a steam locomotive with the engineer and fireman.

Because these trains were so long, the crew consisted of the three men in the engine, the Conductor (boss), a Flagman and a Middleman also called a Trainman.

Radios were not yet used in the 1950s and there were still passenger trains. Passenger trains were on time schedules and had precedence over freight rains. Freights had to get off the mainline to let a numbered train pass. Generally odd number trains headed north, and even numbered trains went South.

SHORT-LINE RAILROADS

There are many short-line railroads throughout the U.S.A. and the world. I'm familiar with one in upstate New York. As a look at the map will show, the main line from New York City to Chicago, follows the water-level route. Up the Hudson River, west following the Erie Canal, to Buffalo and thence further west. This is the heaviest and fastest rail with several branch lines joining with the main line. In the Finger Lakes region, with routes between the legs, rails ran North and South. Roads connect to Lake Ontario to the north.

"YARD GOAT" RENSSELAER COACH YARDS #950....
PAUL HILL   11 7-10

A significant coal road, the Delaware and Hudson, ran from mines in Pennsylvania to Canada. During World War II, in a secret operation, it operated for the army. Uranium ore, pitchblende, was moved from the Adirondacks to Tennessee. This was a part of project 'Manhattan.'

It was allowed to operate during fire season by the army but only if they used first-ever diesel-electric locomotive, thus, not using dangerous steam engines.

Another line, with only 14 miles of track, served quarries to provide ballast rock for rail beds.

The Herkimer, Johnston and Gloversville Railroad perhaps had the longest name. They would meet with the giant New York central, and comment, "Our Railroad may not be as long as yours, but it's just as wide."

Having no passenger service or timetable the H.J.G. line was run rather casually. The only locomotive crew would stop at a stream to go fishing, keeping the catch in the water tender. They also didn't mind picking up groceries for neighbors.

There were narrow gauge railways throughout the Adirondacks. They were replaced by better roads and heavier trucks.

The several lumbering and mining steam engines have disappeared. Except, perhaps, to the delight of children and grandpas, a steam road still serves Santa's workshop on Mount Marcy. Children love the North Pole.

MISCIBILITY (Misi-bil-it-ie)

If you pour a pint of water in a glass and add a pint of alcohol you will not get a quart of mixture. This due to the property known as miscibility. My friend, a physicist, measured the angle of the molecule of water to be 140.° This allows the carbohydrate alcohol molecule to mix in an entanglement taking less space.

For instance, if you had an oil drum full of bowling balls you could add several shovels full of marbles and say it's full. Then, take a bucket of sand and fill the space between the marbles and bowling balls. And then add gallons of water. Now, it's really full. That's how miscibility works.

A bartender will always prepare a mixed drink starting with ice cubes an adding the booze. Then, he or she will fill the glass with mixer.

Ask the barkeep to reverse the order. Pouring in the alcohol last will require much more than a shot glass and be unprofitable.

Anti-freeze for autos, originally alcohol, now glycol, is miscible. Drawing samples from the top or bottom the mixture will have the same distribution. Those so-called freeze out plugs don't work. Because they are really a feature used in the foundry to clean out the core and sand from the casting.

While on this subject, look at grandpa freezing his applejack to get at the center of a frozen barrel, for the 'good stuff.'

If you experiment it won't work. When you study latent heat, the entire mass solidifies or melts as a complete mass.

Place a large, solid ice cube into a glass and let it overflow. With water at the rim and the portion of the ice cube above will the water go down or overflow?

You can stop repeating legends and prove these tales to yourself.

THE VICTORY SHIP

I went to sea as a young man, on a freighter, the Buckeye Express. She was a Victory Ship, built in 1943, at Portland, Oregon. One of perhaps 600 in her class. This vessel belonged to the U.S. Army. Victory ships differed from the more numerous Liberty ships launched during World War II.

Liberties were styled after the most common cargo ships of World War I. They had riveted halls, and reciprocating steam engines. They were slow and operated in convoys, rarely at 12 knots. The Victory that I sailed on was different.

A Victory had a hollow-ground bow and 6500 shaft horsepower, rather than the usual 6000. She could make 18 plus knots in cold water. Being a geared steam turbine, the military sought to use them to keep up with combat vessels.

As a practice, ships that were mothballed after the war were reactivated and operated, under contract, by civilian ship operators, and crewed by merchant marine seamen. We carried military cargo to wherever the government agents directed. We also got mine and dangerous waters pay.

This five-hatch cargo carrier was to be kept clean, for possible conversion into a troop carrier, as she had evaporators capable of making many tons of fresh water daily. The lower holds of number five hatches were fitted as tanks, for extended range. Spaces in the stern could become quarters for an armed gunnery group.

As used while I was on board, the required manning was 38 crew. When at sea, the deck watch was three: a licensed mate, an able-bodied seaman and an ordinary seaman.

The engine room also had three, an Assistant Engineer, a Fireman/Water-Tender and an Oiler. The Captain, First Mate, and Chief Engineer were not watch standers. One radioman, and the steward's dayworkers, completed the entire compliment.

This same ship, KX, FW, when operated by the Navy, required 205 officers and sailors. Let me explain.

If the navy or army mans the same vessel, for combat use, then it requires many additional personnel. The first one on board would be a communications officer with a crypto machine. His radio crew consists of four, replacing the single civilian radioman. Lookout and signal-men increase the men needed for the bridge watch. Regulations state that the 50th person be a medical officer. A sickbay under combat condition, means a staff of pharmacists mates.

The storekeepers, maintenance staff, facilities for heads (restrooms), mess galleys and dining areas also increase. Wartime ships, carrying gun crews, require a combat center manned 24/7.

This 1943 postage stamp features a Liberty Ship unloading cargo.

In forward areas there will not be contract cargo handlers. So, ships must have their own winch drivers.

The increase in crew requires additional lifeboats and rafts, necessitating many Coxswains and boatswain's mates. Ships in combat zones are not refueled. They carry enough bunkers for a round trip. Therefore, subtracting from tonnage being delivered. This requires additional ships, with their hundreds of sailors.

You can see how the manning of combat ships is like growing mushrooms. Imagine a dozen merchant ships in a port, like Lisbon, Portugal. They would have shore time for around half their crews, say about three or 400 seamen in town. Then a Navy carrier anchors, with thousands of white hats on liberty. Dont forget carriers sail with destroyers, cruisers, and replenishment ships in their company.

Citizens – "Look out for your wives and daughters!"

The officers and men of the Merchant Marine, by their devotion to duty in the face of enemy action, as well as natural dangers of the sea, have brought us the tools to finish the job. Their contribution to final victory will be long remembered. – General Dwight D. Eisenhower on National Maritime Day, 1945.

The North Atlantic in January by Paul Hill. Watercolor

BLUE SMOKE AND MIRRORS

The ship I sailed on belonged to the U.S. Army, it was a Victory ship, built in 1943. The Army keeps selected ships for ready use by having them manned and operated by civilian crews. So, there I was, learning to be a navigator in the Merchant Marine.

This ship, like many built during WWII, was equipped with any gear that was available and hurriedly launched to win the war.

For instance, this ship was fitted with an electrical system run on direct current, 120 volts (like railway trains). In later years, there was add a single 120 volt, alternating current line. 120 VDC was supplied to be Gyro Navigation System and the radio room to power modern equipment.

The cabin I was assigned was directly below the radio shack. As such, the many cables to and from the space above were run close to the overhead in a cable tray above my head. This provided an unusual opportunity.

It has been a standing order that no personal radio antennas were to be run on a ship. Also, in the 1950s, before satellite systems radiotelephones were in use. Operating on VHF frequencies, they provided line of sight connection only. Therefore, the radiotelephone was not used a great distance from shore stations. The radio officer provided communications over low frequencies.

'Sparks' was a highly qualified radioman, and his set-up was two special circuits. He ran the alternating AC for the circuit to the

Here I am as a salty young man
contemplating life and days.

captain's stateroom. This was a special arrangement for several reasons.

In those days stereo music players were introduced. Our skipper loved classical music. He bought a very expensive stereo system. As he planned to use it at home when he retired. The set needed 120 volts alternating current. This meant the only radio equipment on the ship that was civilian-type receivers, were the captains and my little Hallicrafter S38 desktop receiver. So much for power circuits.

The radioman on our ship was super. He could listen to the telegraph on two frequencies while typing and still answer a question. He sailed the Pacific, before and during WWII. Not only a professional but also a ham radio operator. He contacted my brother, also a ham, during the holidays using a special rig, made by him, for my little receiver.

Since the many cables from the radio shack above came through the overhead of my cabin, a special antenna connection was made. Using antenna could be used by me under strict conditions. He rigged a neon bulb the size of a peanut, held by a paperclip, to the radio antenna lead. This antenna was used as a receiver when listening. But when the transmit button on the handset was pressed 1,000 watts of power was sent out of the antenna. The neon bulb would glow.

This unusual connection allowed our crew to listen to the Walcot/Charles boxing championship fight in the mid-Atlantic. But – dont forget to disconnect from the telephone lead when making landfall or you will fry your receiver.

Things went on fine for weeks and months. Then our skipper was to take his 30-day annual leave. A relief captain from the home office would take his place temporarily. I was cautioned by the mates to be careful what is said to this temporary captain. We topped off our tanks at Long Beach and sailed for Japan.

He heard of the use of the telephone antenna and thought it was a clever idea. He had the stereo system in the captain's quarters rigged to be the same, with all the cautionary remarks. His choice of listening pleasure was less classic and was heard in the wheelhouse above. We sailed for a week and a half to Yokohama.

The pilot station tried to contact our ship KXFW they called on the telephone frequency. The music from below stopped – disaster struck. He had forgotten to disconnect the music radio from the telephone antenna. The full force of the telephone transmitter went into the radio receiver.

Of course, when the docking took place the damage to the radio was assessed. Fortunately, there are expert electronic services available at Tokyo Bay. Our agent had technicians take the set to be repaired. They promised to return the set before we sailed in a few days.

Ships are busiest when in port and our routines necessarily took precedence. On the day of our departure the repaired stereo system was delivered.

Let me explain, on a merchant ship, the officers' dining room is called the 'Saloon, meaning the finest space on the ship and used for business when in port. Also, let me remind you the whole of the ship is wired in direct current. Not alternating current.

So, the ever-smiling Japanese technicians delivered the set. "Well," said the relief skipper, "let's see how well you did." He plugged the repaired set into a nearby DC outlet. A cloud of blue smoke and a loud groan was heard. The set was totaled. As we were ready to sail nothing could be done.

Upon leaving Japan, we dropped the pilot and took departure. Setting the sea watches, standing in the chartroom, I was told by the captain, "I'm going to keep that set. When we get to San Francisco, I'll get a new one for your captain."

SHRIMP BOATS ARE A-COMING

So the song goes. It popularized an important segment of life here in the Southern coastal waters. We lived within walking distance of the harbor where they berthed. The boats showed up following the seasons, usually around Christmas here at Cape Canaveral.

Dozens of shrimpers crowded the harbor, tied up three or four deep. When viewing the moorings, wed be amazed at the jockeying of a boat from their midst and left to go dragging.

Fuel and ice were taken aboard when leaving at dawn and they went to the fish house when offloading late in the day.

The boats we remember were family-owned, many with wives and children. Their years' work began in the Carolinas and went down and around Florida, to the Gulf Coast, and then to Yucatan. At last, with bare poles, they returned north via the intercoastal waterway.

Shrimping had always been a part of the seafood industry. But it soared after WWII. A classic design prevailed for half a century. The marriage of two features worked very well. That was a wooden hull of southern pine *and* a General Motors 671 diesel engine.

This engine was a six-cylinder, two-cycle design with a vane-type supercharger and top-mounted exhaust valve. It was simple and reliable. Millions of them used from greyhound buses to trucks and heavy equipment. Many 671s were rebuilt into marine versions,

The hulls were of proven design. Many made without plans but just made like the others. The design was very wide for its length. This provided a big workspace aft. The engine was well forward, followed aft by the mast and winches and the hatch above the hold below.

What made this boat a shrimper was its rig. That is the mast with its standing rigging and immoveable cables and the fishing gear with net doors and lines.

It took a good crewman to work these rigs without mishaps.

The catching of shrimp is an art. It has been said, "Like gold, shrimp are where you find them."

As the sunlight penetrates only 300 feet into the seagrass, it does not grow far offshore. Shrimp stay buried in the sand during the day. It is illegal to shrimp at night. To avoid overfishing the industry is well managed by regulations patrols and heavy fines.

The trawl net may be one large affair dragged off the stern or two rigs right and left. When properly deployed the net forms a large open bag – its front taking a rectangular shape. Floats along the top, chain across the bottom.

Doors specially designed keep the 'purse opened as it is pulled over the sea's bed.

Periodically, as dictated by experience or guess, the underwater rig is hoisted onto the shrimp boat's rear deck. The catch, when emptied contains a variety of desired seafood and trash. The crew must work to separate and pack the catch in ice layers in the hold. It's disappointing to return to the dock empty-handed and expensive. Fishermen need a lot of luck.

Shrimpers usually run at a constant depth of 80 to 150 feet. Because the surf and seas on the parallel course make the boat constantly roll. A device has been introduced called a roll dampener. Much like the shock absorbers on your car they cut down the rolling action. Simple in design they are triangular steel plates two feet or so across. Hung from the outermost boom they are submerged by diving down without resistance and causing drag

when pulled up. So, a more comfortable ride is produced. Sailors call them flapper stoppers.

Well, the shrimping industry has been taken out of the hands of family-owner-operators. Gone are the mom and pops. Corporations have taken over. Large, steel-hulled shrimpers have three, more powerful Caterpillar diesels. Union crews, living in air-conditioned quarters, work all year round. The catch is quick-frozen and packaged for market and stored under refrigeration.

After hunting in foreign seas, the shrimp are delivered fresh off the boat.

'Ranch-grown shrimp and prawns are another factor affecting the fishing industry.

NOT A U-2

Everybody knows what a U2 aircraft is. The spy plane that was the cause of a great political crisis when we lost one inside Russia. Well, I'm here and telling you my story of flying a U-1 aircraft.

The designation of U-1was used by the U.S.A.F. for the light, single-engine Cessna 195 airplane. I loved it. It had the feel and performance of heavy aircraft, not like most light planes that use four- and six-cylinder flat engines.

The all-metal four-seat craft was a taildragger with Cessna spring-steel landing gear.

Cesna 195 (U-1)

Also, the high wing was not fitted with a lift strut, that diagonal brace used on most small aircraft.

The cabin was roomy and comfortable, with good visibility and auto-type doors. Cruise speeds reviling the 1930 transports that were made for rapid trips.

Equipped with an engine-driven generator the ship had a full instrument panel permitting IFR flying. The radio and navigation installations were the latest for 1958.

In every way a business ship, the powerplant was a Jacobs radial engine with an automatic propeller. This meant the engine controls were throttle-prop mixture. It is flown like heavy aircraft.

A curious feature was the crosswind landing gear which centered doing landings. It was a delight upon landing, but a pain when parking inside a hangar, often requiring several attempts.

Later designed aircraft have much better performance and features. But I grew up in the taildraggers and this was one of the best.

FEEBLE EXCUSES

I taught at a local high school for 11 years. I liked being with the students and they generally behaved as expected. One problem, we resolved in a unique way.

Being tardy or absent could generate longwinded and emotionally expressed excuses. Especially when interrupting a lecture or test. The reality in some of these tales should be published, and a few of the excuse presenters would earn Oscars.

So, I took a little time to organize a solution. With chalk in hand, we listed many common feeble excuses.

The list went like this:

1. My feeble had a flat tire.
2. My feeble ran out of gas.
3. The feeble driver was late.
4. I spilled feeble on my clothes, etc.
5. Our pet feeble round out of the yard.
6. I had to get feeble money.

The original list on the chalkboard numbered about 15 or 20, but you get the idea. In operation, the students need only mention the number for their excuse to be recorded, and not have to say more.

Students were not allowed to repeat numbers but must exhaust the list.

When a unique excuse came by, it was referred to the Rules Committee. They were the last row in the class – they were the most frequent users and the most creative.

In marking my attendance records, the Assistant Principal, the Dean of Instruction, saw these numbers in my notations. Upon explanation, he thought it would not catch on.

PATENTS AND COPYRIGHTS

Many people think that having a patent or copyright makes them rich and that the government will protect their intellectual properties. Well, there is a lot to be explained on this subject.

I was fortunate to be in the military with a buddy and lifelong friend. We both volunteered for the draft to get our obligation over with and pursue our careers. He had graduated from Villanova and worked for IBM and later, Pratt and Whitney as a patent attorney.

I grew up in my father's machine shop and worked on many development projects with consulting firms. Over time, our visits left us with an understanding of his work.

For openers, the Patent Office is only a registry, and the purpose of patents is to be a benefit to the nation. The process, when followed, leaves a regular trail.

Proof of date of conception, made as a public announcement, may be helpful later. The applications for letters of patent also provide a reasonable length of time to develop a product. By this plan, the notation, 'Patent Applied For,' claim-jumper is avoided.

Because the final granting of a patent number will accurately and precisely describe what is being registered, all new patterns are published in a bi-weekly journal.

Specific personnel read each disclosure to see if there's anything of interest to their companies. If they can accomplish the same invention in a different way and avoid paying royalties, they will.

Robert Goddard developed systems and components for his many rockets. His patents were put in the journal. Few people were interested. But the German, Wernher von Braun, was. International copyrights and patents are enforced by their owners, not the Patent office.

Eli Whitney made his fortune manufacturing firearms and is to be remembered as the father of mass production, having introduced the interchangeability of parts. Instead, he invented the cotton gin. He patented it and spent years chasing copiers. He spent his fortune, and his wife's, trying to collect royalties.

But the longest case involving patents was Ford versus Selden. Some jackass issued Seldon a patent for chemically driven carriages. Ford had few resources, so he used delaying tactics. He engaged lawyers, paying them with stock, as his enterprise grew. Finally, hugely successful, with the largest payroll in Michigan he would shut down if Seldom prevailed. That decision was in Ford's favor.

The business of trademarks, copyrights, and intellectual properties requires the owner to pursue and enforce their rights.

My friend is one of the few patent attorneys who flies down to Washington in a business jet. Being allowed to make their own searches and presentations directly for patent assignment.

He says the government fees amount to $37 in 1962. He offered to get me patents, at little or no cost.

There is an industry of taking money from people with ideas and running the process, with little rewards. He recommended I do not go that route.

If you have an invention, make presentations to interested companies. When offered an adequate agreement take it.

The same idea holds for intangibles like written poems, music, novels, etc. Big money can use their lawyers to tie things up for years period thereby preventing other actions.

Infringement suits are another tactic. For instance, Coke is a property, the word Jeep, capitalized, belongs to Chrysler.

In conclusion, would-be inventors can save time, money, and heartaches by offering ideas to those who use them. Take the money and run.

In today's world, patents and copyrights are ignored. Consumer goods are made offshore, requiring international suits. It is common for companies to disappear and only the lawyers profit.

I WANTED WINGS

As schoolboys during World War II. My friends and I were fans of aviation, especially combat planes and their pilots. Movies, books, comic strips were devoured by me and my buddies. Building model aircraft from kids of balsa wood and paper covered, was a great hobby. Powered by rubber bands, their short flights, fueled our imaginations. Through high school, we were Civil Air Patrol Cadets.

Upon graduating high school there was the universal military training act. You could join some active or reserve outfit or be drafted. Many young men chose to volunteer to be drafted and get one's obligation over with. I was directed to have a physical exam and be tested, for assignment. As it turned out, I was qualified for pilot training in the Air Force. I was jubilant. Overjoyed!

Preflight at Langley Air Force Base in Texas went quickly. The short weeks were devoted to life in the military, issuing uniforms, marching, military customs and courtesies, etc. But no aircraft.

There were several primary flight training bases. I went with a group to Stallings Air Force Base in Kinston, North Carolina.

In the initial meetings the standard message was, "Only one in four of you will be getting wings upon completion of your pilot training."

Of course, I would be one of the few the proud. And later years I met many men who were part of the 75% that did not get their wings

I should mention some aspects of my military involvement. First, is the enlistment contract. As in any contract, parties agree to the particulars. Second, the United States Air Force does not train pilots.

My program was a contract flight school run by the Servair Air Corporation. Four students were assigned to each civilian flight instructor. The school graduated four classes a year. The operation was called 'The Tiger Program.

Of the student body of 600, there were OCS graduates, already officers, and foreign flight students, some having wings. They came under the mutual defense alliance program of NATO.

Except for a handful of regular Air Force TAC officers, everything was accomplished by civilian contractors including base housing, mess, aircraft maintenance, sick bay, air traffic, and all other civilian organization functions.

Flight training began immediately with instruction split between classrooms and flight. The first ten hours of flying was in a two-seat taildragger set up for dual instruction.

The piper PA-18 was a militarized 'super cub. It was not the classic piper cub. With a powerful engine with a full cowl and gas tanks in the wings. They had bucket seats, parachutes, and radios.

I did not have a problem since having some time in light planes while in the C.A.P qualified me for a C.A.A., and now F.A.A., student pilots license.

Within a few weeks we were introduced to the AT–6C 'Texan,' built by North American (come Rockwell). Most of us were surprised and delighted that we should get our hands on this famous World War II advanced combat aircraft. Especially after so few hours being student pilots.

The AT-6C is a heavy, big, tail dragger with a radial engine. At 6050 pounds it was no light aircraft. The size most World War II fighter planes, it was used as a combat trainer. Thus, it was rugged, requiring a lot of horsepower. The radial engine was unlike any small aircraft powerplant.

The Wright nine-cylinder radial was like the power installed in transports. Rated at 695 horsepower, each piston developed 77 HP. Being fitted with an automatic propeller, operating the engine was a challenge. In practice, radial had a set of controls for each engine, mounted on a control quadrant.

Three levers with knobs were marked T for throttle, 'P for propeller, and 'M for mixture

With reference to engine instruments as follows. Your RPMs were set by the prop control. It would not vart during maneuvers. The throttle control was adjusted for the manifold pressure desired.

The mixture control was used to regulate the use of the high-octane aviation gasoline.

There is no oil stored in the engine. Called a dry sump engine, oil is sent through an oil cooler to be kept in an oil reservoir. Before circulating under pressure throughout the engine. As used, this engine used 30-gallons an hour. At the economical cruise maybe 20-gallons per hour.

Flying this ship introduced us to retractable landing gear and flaps. A tough aircraft, the maneuvers we practice were unlike civilian pilot training.

As taught in most flight schools coordinated flight meant keeping upright and never sliding or upside down. These planes were stressed for six positive and three negative Gs. Of course, the students and instructors must also be able to withstand such physical stress.

Acrobatics were taught. Being a nice pilot and always flying like an airline or a business pilot will get you killed in combat. We were taught to fly uncoordinated. That is "stumble" through the air. Recovering from unusual altitudes was a program few experienced. This flight program went beyond the requirements for a private commercial and instrument rating.

Students were sent out solo, to practice stalls, loops, spins, and other tactics in preparation for evaluation tests. Our practice areas where over the great dismal swamp in northeastern North Carolina. The accelerated program included cross country flying, radio navigation, and instrument blind flying, 'under the hood. This life was very demanding. Every week several students quit or were 'washed out. Several were killed

The orders came from air training command headquarters. The TIGER program was to be canceled. Cadets with 65% of their program completed would be allowed to finish. I was at 72%. No exceptions were to be made

At the meeting there were about 100 cadets affected. We were advised of our options. Since this was in the Air Force decision, a 'convenience of the

Art can sometimes capture moments that mere photography cannot. Angel Falls, Venezuela.

military, honorable discharge could be had. As I had enlisted for purposes of pilot training. I could go home. But then there would be the draft.

I, along with others, agreed to extend our service for the time required to satisfy our selective service obligation.

Now the question was, 'What to do with this group of aviation cadets? I wound up being sent to Sampson Air Force Base in Western New York's finger lakes region. Not having any career specialty, I was assigned to the special services squadron. 'Special services means that you do anything that nobody else does.

A VERY SPECIAL AGENT

My brothers and I were caught up in the events of World War II. We built model airplanes from balsa wood kits that were paper-covered and powered by rubber bands. Fighter pilot aces were our heroes.

At night, in the darkness, we played commandos. As American boys know, these special forces are now called Rangers.

Raiding parties in uniforms are to be treated as prisoners of war. Special agents on clandestine missions in civilian clothes are spies. Sneaking about silently in shadows was great fun. Several movies and comic strips were of particular interest.

Smiling Jack by Zach Mosley depicted aircraft and details of them. A character in the comic strip was 'Fatso. He and his wife had twins in a carriage named Dot and Dash. The babies spoke in Morse code.

We deciphered the messages such as 'Buy bonds, or 'Donate scrap metal. My brothers and friends practiced and flashed coded messages at night. I used this skill years later.

I fantasized in my mind how I would become an agent. Parachuting into France, I'd join the underground. I'd make a radio winding by winding wire around a saltbox and sending messages to headquarters, helping win the war.

Nowadays, amateur radio operators do not use code or make their own sets. Model airplanes are ready to fly out of the box. The once popular RadioShack stores are gone.

As a young man, I shipped out in the Merchant Marine. I needed to qualify as a Signalman to stand a bridge watch.

The several ways of communicating at sea included Morse code. Not only by radio sound but also signal lights, semaphore, flags, steam whistles, and yardarm banners.

But the use of the international code of signals transcends languages.

THE PRISONER CHASE

As mentioned earlier, I was placed in a special service squadron at a basic training base. I had volunteered to serve a two-year obligation. The prospect of going to a tech school was out of the question for two reasons. First, the Air Force was reducing in size. Second. I was a short timer.

The special service outfit took care of all the things nobody else did. The rapid succession of duties was engaged in an interesting. I served as a courier for registered documents, hand-carrying papers that weren't to be sent through normal distribution, such as lawyers' briefs, contractor bids, etc. (No plans for automatic atomic bombs.) I served the Base Library, hospital medical library, base hobby shop, and others.

My classification was Administrative Career Specialist. One assignment deserves special attention – I was chosen to go on a prisoner chase.

A not-unheard-of scenario would go something like this. A recruit would not like military life and would go home. Afterward a while, his folks would wonder how long he should be away from training. So, he takes off and heads for Florida or California. Out of money and hungry, he turns himself into the authorities.

Well, the air police have their hands full doing police work at the base. So, they requested assistance for these non-violent absentees. Arrangements were made to escort them back to base.

They cut orders for me to go to Florida and return with two' boots being held at the Orlando Air Force Base. Given tickets and vouchers, I traveled from Geneva, New York, to Wayne junction on the Lehigh Valley Railroad.

In 1955 they were still using steam engines and wooden coaches. Dinner was linen and crystal served by an attendant in his starched white cotton jacket. The lamb chops with panties and mint jelly were delectable.

We changed to the Atlantic Coast Line to a Pullman car in Washington, DC. The end of the terminal was a huge glass wall showing the capital lighted in the night. The Porter turned down our berths. It was a long day. The sheets felt terrific. It didn't take a minute to go to sleep. I didn't even notice the train pulling out.

I say we, for other airmen were traveling south on other assignments.

Rising early, dressed in our uniforms, we heard the tones of the chimes announcing breakfast in the dining room. This train was a late model, air-conditioned set of cars pulled by a diesel-electric locomotive. We were speeding through the rice fields of South Carolina. Arriving at Church Street Station, we were carried to Orland Air Force Base, off East Colonial Drive just North of Orlando airport.

At that time, this small base of about 600 personnel was headquarters for aerial photo mapping selected for the nearby golf courses. We checked in did a lot of paperwork and were given a short tour of central Orlando. It was city beautiful.

I wasn't taking the place of a badged and armed air police. So, that there was complete understanding, I met my two charges and explained things and what to expect. I didn't even have handcuffs or any means of restraint. This was to be an uneventful trip.

I told them, "Enlistees are not really in the military until after their 89th day. You are not yet a deserter because you choose to return. When you are back in your squadron, you'll meet your commander. He will ask you if you choose an 'Article One,' or a 'summary court-martial. The difference is as follows. An Article One used to be called company punishment. When you complete basic, there will be no record following you. Should you choose a Summary, you will be tried by the same officer, and they will make a record of your offense. Had you pled "Not guilty," in hope of being freed.

"You see, under the universal code of military justice, persons are charged in two parts — the charges such as 'robbery or 'murder. The Specification defines the charges like 'possession of goods or reasons that mitigate the charge. You plead guilty to both, innocent to both, or split.

"My recommendation is to plead guilty to both 'Charge and Specification.'

"I'll also tell you what your punishment might be. You'll be returned to complete your basic training, given some extra duty, and loss of pay for a specific amount. The amount will be the cost of my trip to get you and the fare for you to return.

"You see, there are no budgeted funds for airplane tickets not ordered."

That trip from Orlando was out of the Orlando airport and an Eastern Airlines Lockheed Constellation to New York. Then New York central railroad to Geneva."

The two young men were checked into the stockade to be processed and returned to training. I never saw them again.

THE DANCE COMPANY

I grew up in a large family of blue-collar workers. Down by the tracks, one became street smart and tough.

Circumstances and the years took us far away, with a family of two girls and a boy. My wife and I wanted all the best for our children, and fate involved us in dance.

Let me first say the people involved in dance are the finest of fine folks. Dedication to such a demanding life leaves no time for waywardness. The teachers, staff, students are all of one mind, and few get into trouble.

My wife and I made a trade-off with the studio owners. Without charge, our two girls took tap, ballet, and modern dance lessons. For boys, by custom, tuition is waived because they much sought after as students.

As a skilled handyman, I helped around the studio wherever needed.

Two entities were operating under the same roof. The dance company was a tax-free cultural group, and the business for profit was the dance studio.

Our involvement centered on two duties. I built the stage properties before performances. Fortunately, there was little need for scenery and furnishings. I quickly made simple sets of donated materials with my own tools. Few were durable items, and all were discarded afterward due to lack of space.

Before performances, my wife was busy with costume fitting and alterations. During the several days of actual performances, Shirley did makeup and hairstyling. I served as stagehand and cameraman. This was all new to me.

We, another dad, and I, operated two video cameras. We set one camera to take in the entire performance end to end. The music track was seamless, and everything the audience saw was recorded. The second camera was flexible and used a zoom lens.

In those days, videotaping was still new. Our tapes served two purposes. Videos of full performances were made available to family and friends at a small fee. More important, as an instructional tool and résumé.

At the cast party, after performances, individual students transcribed their own performances to be used when needed for career purposes. Those pizza parties allowed the building of lifelong friendships.

A few happenings come to mind. Like the time we burned down New York. The scene required a metal silhouette of a skyline, fitted into a spotlight, as background. The metal slide did not arrive for dress rehearsal. So, being the handyman, I made a slide out of a flat and beverage can. We tested it for a short time, and it worked.

You see, the professional slide is made of steel, but the homemade unit was made from an aluminum can that melts at a high temperature. That was a show to remember.

One of our graduates became a well-known professional. He moved to New York and, over the years, was quite successful.

A banner performance was made at Brevard Community College, dancing with a world-famous ballerina. The house was packed. I should mention one group of younger dancers delighted big grand folks with 'Up, Up and Away.' They released balloons that rose into the darkness above the stage.

We stagehands expected the balloons to lose their buoyancy overnight to be swept up. A lone balloon floated out from the stage above the audience. It lingered and moved slowly in different directions, distracting everyone.

All eyes were on the wondering toy high overhead. After the world-class ballet was completed, the renowned professionals left with an explosive, "Peasants!"

These many years later, we meet people around town and in church who reminisce. And as I said, you couldn't meet nicer people.

PATTY AND I

Anthony Hill, my father came into possession of a so-called summer home. It was a two-car garage made into a living space. Originally with an outhouse and dug well, there was no electricity, and it was on a dirt road.

The redeeming factor was that it was riverfront property. Over the years especially after World War II, a full-width screen porch, dock, plumbing, and electric were added.

Our camp was located between two farms, and among a half dozen similar camps. They provided escape from the city's hot summers, and a break in the routine imposed by work and the school year.

The farm to the north was quite prosperous, with some good bottom land and a well-managed dairy herd. The Dugans did very well in life.

The Baileys to the South would struggle with poor land, a water-soaked bog along the river, not suited for development. They lived on the edge of property. Living in the house, in need of repairs, were parents, children, and grandparents.

My family were all gun owners and sportsmen. Our menfolk not only hunted and fished but were involved in the black powder arms events, so popular in this region. We needed little excuse to target shoot or go woodchuck hunting.

The farms encouraged getting rid of the ground squirrels. Always asking permission, we'd be waved on with a "Be sure to close the gates."

One of us had a 22 Hornet rifle with a scope on it. The shot went flat and fast. We were all were good shots. When we bagged a couple of woodchucks, we turned them over to the Baileys who were happy to get them. And they blessed us.

The Hill brothers: from Left to Right:
Bud Hill. Bill Hill, Paul Hill, Fred Hill, Bob Hill

My brothers, Fred, Bud, Bill, and I enjoyed helping on the farms on the days they needed to make hay and silage.

These were times before bailing hay and making silage was done with horses. We learned that hay was to be cut, tended, and stored in a manner to preserve its food value. Straw was the stalks of grain such as oats used for bedding and not animal food. Loose hay was hand fed by pitchfork and stored in hayloft, above the cow stalls.Silage was whole cornstalks cut up in pieces a few inches long and blown into silos adjacent to the cow barn. This process usually involved neighbors swapping work with neighbors. A large Oliver tractor with a power take off pulley on its side used a long flat belt, always with a twist in it, to drive the silage cutter. A powerful blower blew the chopped corn into a pipe to the top of a silo.

A farmhand was inside the silo packing the corn. He wore a large pouch of salt suspend over his neck. He was, in effect, making kraut; preserving forage for the herd to eat during the winter. Cows had to be weaned on to this different diet. People could taste the difference in the milk.

Patty Duncan and I were schoolchildren and running about the men and machines was great fun. We rode the hay wagons, jumped into the hayloft, and even climbed into the silo being filled with corn.

The worker was swinging the corn about with a pitchfork and we laughed and ducked while being showered with silage.

Disaster struck – the long sharp time of the fork when right through the palm of Patty's right palm.

The operation was stopped. Everyone was greatly concerned.

A shirt was torn into bandages and, keeping pressure on the wound, Patty was rushed to the kitchen.

Mother cleaned and dressed the punctured hand.

Farm families did not rush to the doctors or hospital for every cut or bruise. There was no 911 and they had no phone.

Receiving much tender care, cleaning and changing bandages the hand was soon on the mend. By watching out for infection the redness subsided and Patty went to school with a story to tell.

We parted ways but our families stayed in contact. I went to her wedding and was re-introduced to her. I took her right hand and drew it up and kissed the double scars.

FAKING BUSINESS

My mother came from the old country, before World War I, as a ten-year-old girl. She remembers living in Yonkers, New York and going to Van Cortlandt Park to watch the filming of Jack Holt western films.

She called all the movies, "faking business." The filming business moved to California and took adaptation of the truth with them.

For instance, I have an interest in aviation and can no longer remain silent. Altimeters do not read the height above ground. Also, aircraft engines turn slower than auto engines. Most airline captains are baldheaded and deaf. They usually drive their second car to park at the airport. Few look like Robert Stack.

Fighter aircraft of World War II used 30 caliber machine guns with ammunition much like a deer rifle. Shooting a target several hundred yards away was useless.

The B-17 bomber entered the war late. The RAFs Lancaster heavy bomber did most of the work at night. It was flown by a single pilot, usually a sergeant. An officer/navigator bomber was in charge. Girlfriends were not allowed in the flight areas – not even June Allyson.

The business of uniforms is disturbing. Dress uniforms are stowed, not worn in action. Did you ever see a hat, called a 'Piss cutter, that is shaped like an envelope, worn? Dress hats that were awfully expensive, were not to be lost in a gust of wind.

How actors kept them at the back of their heads is a wonder. Dress code says it should be worn "squared." Officers unlike Colonel Hogan of *Hogan's Heroes* would not be in prison with enlisted men. That show was a laughing matter.

Most troops in combat were young man, not like Telly Savalas. He was in a war film trying to capture a German tank fuel supply. Anyone could've told them German tanks used diesel engines. American tanks had aircraft radial engines and would need high-octane aviation gasoline.

Now, let's hit the trail for westerns. The six-shooter pistol fired six bullets and then needed reloading. U.S. Army uniforms did not look like stage costumes. Even thoroughbred horses cannot do full gallop for several miles.

Wagon trains were oxen or mule drawn freight wagons filled with belongings. Pioneer folks slept under the wagons.

They settled on lands and went to bed exhausted. How did all those men afford to stay night and day in a salon? Did they really toss down straight whiskey? Who's watching the cows?

Now for the 'Whodunits. Shouldn't that be, 'Who did it. Hollywood police only do homicides. Unless you have a dead body, they don't care. Modern films have only plain clothes detectives who stay in the dark squad room or go to the morgue in each chapter.

Cops work a caseload of dozens of crimes and half go unsolved. Unlike Perry Mason, who gets a criminal to confess at a pretrial hearing. And he never once called Hamilton Burger, hamburger...

Years ago, one would sit in the movie theater, in the dark among strangers, and be scared. Shown on television, the monster movies or terror scenes appear on our small screens, at home, and are conveniently interrupted by commercials. The effect is not the same, but, as mother said, "Its all faking business."

A VERY SHORT HISTORY

I was a schoolboy doing World War II and understood the news and events as heard in the conversations among my father and my uncles. The war began in Europe. Hitler seemed to be winning The United States provided goods of war but resisted armed intervention. The Japanese bombed Hawaii and invaded the Philippines. Immediately, Hitler declares war on the USA.

The first response was to defend the Panama Canal. An obvious target for the Japs. But it didn't happen. They went South to capture the oil rich Dutch island of Borneo.

American, outraged, gathered up a huge strike force an invade Morocco. People everywhere hung maps on their walls unstuck colored pushpins in them. There was great confusion, as to the location and pronouncing names of foreign places.

The largest church in our neighborhood, became full on Sundays and added special novena's, Rosaries, and rituals. American flags were in demand and displayed everywhere. Special little flags were shown in the windows of homes of servicemen.

The civilian workforce was working overtime. Although wages were frozen, and no goods were to be bought it became patriotic to buy war bonds. Also, donating to scrap drives, especially metals, was encouraged. Food and gasoline rationing created black markets. Everyone, the entire population, was mobilized.

The area we lived in was heavily industrialized and considered prime military targets. Blackouts were practiced as a precaution. Air raid wardens, part of our civil defense, were issued arm bands and helmets.

Our hometown, being a major rail center, saw many troop trains and heavy use of the steam trains, double-headed, in long strings of cars, pounding up the surrounding hills. A later joke was the three in the morning freight stopped running and the birth rate went down.

World War II has been chronicled in all sorts of media, in many ways. Finally, the many enemies wore themselves out. The nations turned to rebuilding. People cashed in their war bonds. There was no profit. Inflation ate that up.

The war production resulted in a scarcity of civilian goods. People were sick of war.

Collateral damage, that is nonmilitary damage to civilians and property, far exceeded military casualties. Death from disease and starvation was horrific. The war industries still use a percentage of our national resources. We are convinced that we need to defend ourselves at all times.

In conclusion, I recall hearing war is nothing but big business getting bigger.

HARD COAL AND SOFT COAL

Coal was king. America ran on coal up through World War II. Railroads, steamships, factories, power at stations and heat for homes, all depended on that black gold.

Not only used as fuel coal was the source of many products. Dug from open pits or removed by tunneling, massive amounts filled trains every day. The nations appetite seemed insatiable.

There are basically two types of coal: 'hard' coal and 'soft' coal.'

Hard coal, called anthracite, has fewer calories per pound. It was used in home and businesses for heating, being less dirty to handle and slower burning. It was delivered from coal yards to coal bins everywhere.

Soft coal, called bituminous coal, was highly sought by industry and transport. It was dirty and felt greasy to touch. Soft coal contains coal oil and kerosene and burns much hotter. Each ton of soft coal produce more steam, in boilers.

Many mile-long trains of coal cars were pulled by locomotives fitted with tenders filled with the same. Specially built hopper cars had open tops, steel sides with sloped bottoms and chutes with gates, for rapid emptying.

Heavy industry grouped together to benefit each other. Typically, coke plants were near pig iron blast furnaces. Related industries hugged each other. Soft coal was converted into coke by heating in airtight ovens. Vapors went up to make coal gas and heavier liquids precipitated down, to make coal oil and tar.

Coke is almost pure carbon and burns at a higher temperature. Coke, iron ore and limestone used, as a flux, is charged into a blast furnace. Pigs of iron are sent to be made into steel. The slag is used in road material when mixed with tar, called macadam. The coal gas is piped to gas storage tanks and sent neighborhoods for heating and cooking .

This network of coal related industries changed with discovery of oilfields throughout the world. Not only liquid crude oil, but the abundance of natural gas revolutionized the world.

Pipelines connect the oil wells to tanker ships waiting to deliver their precious cargoes to refineries. Crude oil is transformed into hundreds of products. Not only fuels, gasoline, and Lube oils, but everything from cosmetics to pharmaceuticals.

Natural gas is separated and either sent to markets as a vapor or made into liquid fuels, like butane or propane. Pipelines crisscross our nations. Relieving the need for oil tank cars.

Coal hopper cars on railways have been covered and used a as grain carriers. Diesel electric-electric locomotives Can be coupled in tandem and operated by a single engine crew. The railroad workers don't have shovels.

Rather interesting – the big-inch pipeline delivers gas to salt mines in Western New York. The underground caverns, where salt is mined, are as big as several aircraft hangars.

During the summer there is a super abundance of natural gas. Liquification plants put the gas away for its needs in winter period

There is still need for coal. Generating plants are located near the coal fields. Power is sent economically via high-tension lines to market.

Science and technology have provided scrubbers on smokestacks to cut down pollution. Open pit coal fields have replaced many tunneling mines. We all benefit.

Take a scientist to lunch.

HOME SWEET HOME

Comparing houses in cost and construction, many comparisons may lead us to interesting observations.

Houses are built to local standards and building codes are adopted to meet geography and weather.

I am qualified to speak to this subject coming from a large family of building tradesmen and many years of hands-on experience. For now, I'll limit my discourse to comparing Florida home construction to homes build in northern New York.

To begin, unless one can build a home out of pocket, one needs financial help. A lender will only allow us to buy a house within our ability to make payments.

Size and location are major considerations. Frankly speaking, a building will be made pretty much like the one next door. Melbourne Drysdale, the banker, wants to be able to resell the property if a repossession occurs.

There were, for many years, people referred to as 'Woodchucks. In the country, usually on a piece of farmland, a project would be started. Digging a hole for the basement and pouring the footings was first step. Concrete blocks do not say 'cinder blocks, is a start. With a slab and center beam , a floor deck defines the box shaped basement. With tight money, some people move into the space just for the winter. Resources run out, circumstances change – and when driving about one can see these locations.

The footings as mentioned, must be located below the frost line to prevent heaving. Much like a grave, its dog six feet deep.

In Florida, and many coastal locations, houses are built at grade. Digging down to 'undisturbed soil for about 16 inches, for the so-called integral footing feature of the four-inch slab.

Some houses are board framed. Block walls are preferred. Pre-built sets of trusses are held down by an in-place wind beam on the perimeter wall. This provides a clear span within this 'box for partitions to be placed, defining the floor plan. You can remove partitions and change the floor plan. The roof will not fall down.

Florida homes are built without attic or basement storage spaces, leading to problems of clutter and, thus, spawning many thrift stores.

When comparing northern and southern homes, certain considerations came to mind. Building lots prepared for construction must have facilities for water, sewage, and other utilities.

Grubbing in the sandy soil, unlevel terrain or better, hydraulic filled waterfront sites, can be very profitable.

Heating and cooling systems are cheaper here in Florida. Homes are priced – rule of thumb – by living space.

Labor is less costly here in this 'right-to-work state.

Modern material, many plastics, chip-board panels, and synthetic, not asphalt, are highly combustible and release fumes. Storm damage to modern homes makes them less likely to be repaired but, rather, replaced.

Consider – if we are living in the basements without wood framed living quarters above, we are woodchucks.

RAMBLINGS

There are three considerations I want to braid into one yarn. Not that I expect anything to come of them – I just want to express them and dismiss the topics. You can do as you wish with these.

It all began in the 50s. I was building and fixing boats. Although I wasn't a fisherman, I couldn't get away from them. My concern here is trawlers that drag the sea gardens for a living

The practice of catching anything you can in a drag has been around for ages. Boats, like shrimp boats, pull in their nets and dump the catch on the deck. Crews separate the marketable items and dump the trash back overboard.

It used to be boats would take 'bunkers to be processed into fish oil and good fish glue. But products like paint with lead in it and furniture glue are no longer made.

The process of taking 'good seafood and repeatedly trashing the bottom is a disaster. Places like Norway have suffered from excesses of foreign fisheries.

I now want to consider food and nutrition from another direction. There was a naval medical officer named Tom Dooley. His several books were best sellers back in the 50s. He left the Navy and went to southeast Asia as a missionary doctor; much in the image of Albert Schweitzer in Africa.

One of his observations was that natives lived on a diet consisting of homemade noodles. Malnutrition led to lack of energy and health problems. He made a proposal, echoed in commercial fishing press.

Stop throwing the trash fish back! Convert the bunkers into fish meal. Process of maybe four plants located in the Northeast, Gulf Coast, West Coast and Alaska. Specialists could radiant the meal. Radiation would permit the powder to be kept on shelves. When mixed into noodles a better diet would be created.

Universities are developing food products which do not need refrigeration. I've heard of milk being kept on store shelves and freeze-dried foods. Dehydration, smoked food, pickling, and such have long been in existence.

My next consideration is the widespread use of fuel cells. Years ago, the Apollo mission used fuel cells to power space craft.

Fuel cells are limited used today in such applications as remote weather and navigation sites. By installing fuel cells in towns and country sites, fueled by natural gas, overhead transmission lines could be replaced. The vulnerability of our power grid such as after tropical storms is a consideration. Remote locations such as ranches, resorts and the military could be powered by fuel cells requiring fuel storage tanks.

Huge coal-fired power plants and their high-tension transmission systems and transformers can be minimalized.

Of course, this will be costly and objected to by the existing power interests. But Ive been thinking of these changes for a long time.

CROSLEY AUTOS

There have been small cars offered to the public over the years. But it's a matter of timing and economic cycles. Some makes, such as Austin, were like full sized autos, only shrunk. Made of just as many parts using the same material sand machining, the only difference was less weight.

Another approach to a small car was to simplify design and use standard parts. Full-size cars by large carmakers are characterized by their stylized bodies. The curved, deep drawn bodies are made with expensive dies and they are awfully expensive to prepare.

By stamping many thousands of body parts and using assembly-line techniques, cost per unit is held down. Many small vehicles are characterized by flat panel such as the World War II military jeep. King midget offered two-seaters with Briggs Stratton engines. The one brand that's hung in there the longest was the Crosley.

A neighbor had a Crosley with a two-cylinder, air-cooled motor. It had bungee cord suspension, like a piper cub aircraft.

Once, while carrying a goat to his place, the goat ate the upholstery. As a prank a gang of students lifted the car and placed it in a closed courtyard requiring the fence to be laid down.

During World War II, Crosley developed a unique engine for the Navy. Thousands of Handy Billy gasoline powered pumps were made and widely used. Especially by the Coast Guard.

The engine block consisted of two halves made of stamped steel. It needed no radiator or coolant pump as the water being moved by the centrifugal bilge pump circulated through the engine.

The light weight, four-cylinder, motor had overhead valves driven by camshaft on top. Not using castings, the motor was unique.

It was offered in a completely redesigned Crosley – lightweight and cheap to buy. It was quite easy on fuel. My oldest brother had a used one. He raced in local events and won several because of the handicap standings.

The same engine, because of its power to weight ratio, found its way into a molded plywood boat as an inboard. It was demonstrated at a lake at the New York State Fair in Syracuse. My brothers and I all had promotional rides.

Over the years I have seen several of these boats in use.

The most interesting application of this engine can be found in an airplane.

Famous aircraft designer Mooney designed and built a single-seat small airplane. He proposed the army you special troops to fly the small craft. The Crosley engine used a DC-3 oil cooler as a radiator and had a V-belt drive for the propeller. This model of 'Mooney's Mite' was rejected by the military. Several Crosley-powered planes existed as 'restricted aircraft.

The FAA awarded a certificate to a model fitted with a small aircraft engine. These rare planes show up at airshows.

While working in Orlando and flying some charters, I made friends with a Mite owner. He offered me the use of the plane to help me keep my ratings current.

His generous offer was for naught. I'm a big, tall man and I couldn't fit in the little aircraft.

ALBANY THE CAPITAL

New York City was the first capital of New York State. Then Kingston. The burning of Washington during the 1812 war caused this site to be moved to old Fort Orange in Albany.

It is a beautiful location, and its old-world charm has been preserved.

At the head of State Street, the massive stone building is a blend of architectures. When viewed from a distance it seems to be four-stories high. A huge set of steps and Capitol Park slopes down to the plaza. The front lawn is dominated by a bronze General Philip Sheridan on horseback atop a stone plinth. This adds up to a find location for mass gathering and for photos.

The back of the building has a park. Circular in shape. Tunnels connecting the state buildings nearby lie beneath the park

Technically these are utility tunnels. For, when view from above, the many gabled roofs with dozens of chimneys have no smoke coming out of the many flues.

Steam heat is being piped to these buildings from a huge plant several blocks away. Cable trays for electricity and communications are hidden in the tiles above the pedestrian passages.

When built there were very tall, vaulted ceiling and hallways in the grandest style – stone carvings tiles with hardwood moldings. It adds up to a fine palace being built.

But the demands of our modern age have caused many so-called improvements. Without causing a shut down, and with few interruptions, a well-planned transformation began... and continues.

The very high ceiling hallways were framed with steel are made into two floors. Space at the top, where there were arched ceilings, are used for air conditioning ducts. Some lovely fireplaces are now contained electric boxes. Flues now are used for chases for vertical conduits and pull boxes. Catwalks suspended on top of the cable trays give access to elevators, were installed .

As the elevators near completion, one notices nine buttons allowing access to two basements and sevven floors (six above the main floor). Historic features such as the Rose Staircase have been preserved. Working in the very heavy stone basements, broadcast studios for radio and television are included.

The needs of state government are keeping up with the expanding economics and population. Look into a tour if you visit Albany – especially during the famous Tulip Festival.

NEWSCASTERS

From the Westinghouse building, Norman Brokenshire, the first radio newscaster, spoke. "If you receive our signal, please send us a penny postcard." Since the early days of broadcasting news has been sent to listeners, and later viewers, by journalist who became our source of reports.

When Walter Winchell came on the air, my father cautioned all the family to "Hush up1" Edward Murrow hit the airwaves from London during the blitz. The wail of 100 sirens and thuds of bombs could be heard in the background. World events of great importance hold our attention and filled our conversations.

After the war, radar became television. Soft-spoken Dave Garroway related events and announcements to viewers. Houses all spouted rooftop antennae. Broadcast magazines, such as 60 Minutes, expanded and discuss topics of current interest. Journalist put their reputations and careers on the line when serving the public. Our lives, public and private, were ruled by decisions learned from the news.

The last of these reliable informed this was Walter Cronkite. He came to New York from the Midwest, train himself to speak 120 words a minute, in his authoritative bass voice. He also spelled the end of journalism. Not satisfied with the official accounts, he took to the field to be able to say, "That's the way it is."

There was a series called the Big Story, whereby investigative reporters sought out corruption and crime. Seeking justice, they were given rewards and recognition. Of course, this process was looked upon with distain by certain people.

As a boy during WWII there were no fences anywhere. We ran all over railyards, roundhouse, City Hall, the jail, factories, and ships; you name it. We walked through government buildings, sipping from water coolers. The world was open for inspection. Then things changed.

'No Admittance, 'Keep Out, 'Apply at Office' signs and other measures restricted access to workers and their locations.

News persons must present credentials and are restricted to Newsrooms made comfortable with coffee, desks, typewriters, and telephones. They are not allowed to roam. Prepared news releases are brought from various offices by civil servants. Only closely scrutinized, limited wordings are available. Our government corporations now have a siege mentality and have become defensive.

For instance, 'Ma Bell used to send a single IBM card as a monthly bill, you could watch a mechanic work in the service garage. We are no longer allowed to watch airplanes landing or depart. The public, which supports these great enterprises, is kept apart and not to be involved. How are young people to be motivated to go into these many careers

I've noticed news programs on TV are called shows. The presenters are show people. Perfectly coiffed the handsome readers speed read news items. I guess elocution and enunciation is no longer taught.

Rather than informing the public about critical issues the so-called public information outlets are entertainment and we are not to be alarmed. Too much time is spent on a weather forecast every ten minutes, presented by handsome but less-than-professional, non-official, weather 'professionals.'

Cameramen are made to present the faces of actors, individually then together. This practice might suggest a return to radio.

With the continuous interruptions for commercials, topics change rapidly, and no time is allowed to dwell or expand on any item. Flashes of mini film clips do not make informed citizenry.

The news business has certainly matured. There is no need for scoops. We are being well-fed on predigested news releases and are led to believe we have nothing to worry about.

A LOT OF BULL

Geography dictated the location now and the building of so many towns like this. Natives used the waterways and trod the forest trails. Colonials made amends to Post Roads and the railroads follow the so-called water level routes.

Main street in Watervliet was a classic. Granite blocks for roadways were bordered by thick bluestone slabs for sidewalks. Two-story framed buildings stood on narrow lots, side-by-side, with few alleyways. The stores below allowed the owners to live above, alternating with several rentals in two-story buildings. Most were over a century old, cold water and stove heated.

The scene for our action was in front of the bank where the straight-and-level street rose to the level of the steel-truss bridge which crossed the mainline tracks of the railroads

A Ford model-A truck traveled south on an average summer day. This one-time rack truck with dual rear wheels was modified for a special use.

As built, the sections of the side rocks and their sockets were removed. The usual one-inch flooring was replaced with two-inch boards. Sides and front were also made extra strong. This vehicle was carrying a bull – a half-ton farm animal used to service the dozens of cows on the many nearby dairy farms.

However, it happened, witnesses vary, a near-miss traffic incident caused the screeching brakes and the blare of several horns, followed by loud shouts.

The startled beast reared up and tore off the ring from its nostrils. It immediately departed the truck to survey its surroundings.

The bull was scared. It was in an unfamiliar place. A few vehicles left the same and were a few pedestrians about. A couple of persons from the bank sized of the situation and retreated into the bank. The sight of the large animal raging in the street, with blood from streaming from its nose, caused everyone to take shelter and view from a distance.

The police station in the townhall was but the block away. The town's only patrol car quickly arrived on the scene. Using a handheld, battery-powered bullhorn, the officer made loud dictates over the situation, thereby attracting a lot of people, many of whom hadn't heard anything prior to this.

Consulting the farmer/owner, the cop was about to restore order by shooting the bull. This startled of the animals owner because the bull was his livelihood.

He assured the officer that he would control the nervous beast and the officer took control of the onlookers.

From the bulls point of view, he had only one option – leave the scene. To the north buildings stood shoulder to shoulder lined with people and vehicles. To the south the street ramped up to a black steel bridge with railroad engines moving about and snorting. On the west side was a six-foot-high, chain-link fence. The view was flat tracks the river – an open space. The frightened animal, using his bulky shoulders, paced up and down the length of the fence seeking a weak spot.

The farmer, with helpers, fashioned a noose at the end of the rope; not a thin lariat one saw in in the movies. The rope, as round as a broomstick, was draped at the end of a ten-foot pole.

After several attempts the best was looped about his neck. A crew of three, in the truck body, belayed the rope and strained to finally get the, now tiring, bulls front half into the truck.

The bull now rested in the rear of the truck bed. However, any jerk by the handlers tugging on the line impacted the animal in his groin – the bulls special parts.

A halt was declared my both sides. Nearing exhaustion, the owner made a quick and decisive move. He took a firm hold of the bulls tail and bent it into a u-shape near its base.

The startled, animal sprang into the truck bed. The helpers took a short purchase holding him in normal position for transport.

Within minutes the onlookers left. The traffic returned to normal .

This happened in 1945 when news events overshadowed everything else – atomic bombs, death of FDR, NATO, the United Nations, and other compelling events.

Besides artificial insemination was being introduced by veterinarians.

There were with the huge stud fees for racehorses. But cows, dogs, zoo animals were also to be bred by artificial insemination

No longer does bull meet heifer. So, it's a win for science but a loss for Romantics.

WRITE AND WRONG

Writing can be in many forms. The concept of recording thoughts and recording language in written form takes back several millennia. We have scholars deciphering the writings of ancient times; enabling us to have insight into our distant past.

As humankind populated the world and civilized into diversified nations with cultural concepts and languages, writing needed to be clarified to be useful on a broad scale. Word meanings, spelling, sentence structure have all been modified, and changed over the ages.

Most children learn spoken language from their mothers and begin the study of structured language and writing in school. The word school indicates groupings by ability to learn.

It has been said that the average daily use of a language requires about 3,000 words to function in society.

Some people go way beyond what might be called 'street talk. They seek perfection and operate in very narrow parameters. The average person, such as I, violate spelling, grammar, and other rules every day, yet we function in our world. Teachers and language specialists stay in their ivory covered sanctuaries, trying to keep up with the evolving ways of communicating.

Each year, new and old words in vernaculars are added. Old words or the meanings are brought to our attention. At best, writing and the use of words is fluid, constantly changing process.

Just look at the writings of our founding fathers. The constitution would need many corrections. Better yet, the greatest writing of all time, the Bible, was produced by evangelists of little training.

The survivors of the whaler Essex didnt prosper as their tail was written by Melville. Selkirk was marooned and suffered. But his story was stolen.

Many people have lived well by means of Christs teachings. He never left a written word.

So, we have accepted the idea of literature of living very well off the tales of others.

Native Americans lived to their own satisfaction, relating history by word-of-mouth. Europeans showed up with pen and ink Inc. The rest is history.

Never worrying about dangling participles or split infinitives, shamans held crowds' attention throughout the ages.

I once saw a copy of Lincolns Gettysburg address corrected and graded by some screwball professor. How dare he!

My advice to everyone with a tale to tell a simple. It's your story – don't let them steal it. Make them stay in their libraries in classrooms while others are out in the real world.

THE MASKED MAN

Turn on the TV and there was Tim Allen of Home-Improvement talking across the fence to his neighbor. You never see the actors face. Someone decided, and the actor agreed, not to identify the neighbor.

Of course, there have been masked man of great notoriety in real life. The reasons for not wanting to consider reveal one's identity may differ.

Train and bank robbers use the popular bandanna to cover their little faces, such as the James brothers, Frank and Jessie. Recent desperados use the modern ski mask or nylon stockings as disguises.

All who wish not to be identified are not bad persons. In fact, and in fiction, there are good guys who wear a mask, including specialty togs, thereby being uniquely identified. There was only one Superman, one Batman and one Robin. Superheroes stay in their own territories Gotham or Metropolis.

However, in speaking of the masked man, no one can challenge the reference to the Lone Ranger. The mystic built up around this character is curious in many ways. When popularized on radio, listeners used their imagination scenic details. The televised ranger and Tonto are revealed in splendor. Viewers must accept the established storyline. As a youngster, I was a believer and was delighted to be a fan.

Now as an elder, I want answers to many questions regarding the 'rider of the open range. Does he sleep with the mask on?

He rides a white horse called Silver. Why not call him 'Whitey, or 'Snowflake?

Are his silver bullets special order or is he a reloader? When going into town incognito, why wear beard and eyeglasses. Take off his mask no one will recognize him.

He often cooperates with the sheriffs. But they never require a proper identification. Does he have a second set of clothes to be laundered for hygienic purposes?

Always sleeping outside of town, never in the bed. People see that Tonto has aged.

These and other curiosities may come up. But hey, if you believe in a talking mouse, anything is possible.

As a parting thought, if Clayton Moore walked in right now, who would recognize him?

THE WARPING TUGBOATS

I grew up beside the busiest river on the East Coast. The port of Albany was strategically located in upstate New York at the top end of 'deep0water shipping, ocean ships met with canal boats from the west and north. Three major railroads met there, and highways radiated. Manufacturing and commerce flourished.

There are dozens of locks situated along the canals. Each lock bracket (or set of locks) is located at a dam, holding the river channel to adapt for shipping. Each dam has its own generating station, spawning industry in nearby towns.

This hub of activity provided a grand number of distractions. Myself and my buddies never lacked for an adventure nearby.

Roaming the railroads from freight yards from roundhouses coach yards and switch towers – no place was fenced in. The workmen knew us and cautioned us.

The bridge tender z the neighbor and let us ride the swing drawbridge. Wed walk miles each day sipping water from faucets. There was a monastery with a pond for us to swim. Or resting on the river. A myriad of adventures.

But the most interesting of all was the ships and boats. The largest grain elevators in the world were here. I'm sure most of the world's pigeons too. In later years, I worked on the railroads, worked as a longshoreman and went to see on a freighter.

The greatest thrill was some rides I had on a tugboat. It happened to be owned by a school friends family. They operated two tugboats, both right in our neighborhood. The large boat was always away with barges on the river and canals. A smaller vessel stayed mostly in the port area assisting ocean ships.

Captain John would tell me to go and tell my mother that well be gone for two hours. I ran back cast off the last line and jumped on the throbbing vessel.

We went to short distance to the busy turning basin. We were underway and nearing the giant ocean ship tied toot the wharves. The

STEAM TUGBOAT   PAUL HILL 2/12

procedure was routine and done in an orderly fashion.

A tugboat like ours was really a floating engine, and a little else. Built in a stout hull built for abuse. Tugs have large slow turning screws and big rudders. Rudder power is necessary for the job at hand.

By a system of signals from the pilot on the bridge, we acknowledge and proceed.

this maneuver was taking place during slack tide with a little current from the river. A light breeze was blowing.

Most ship spend their lives in open seas going along in a straight line. They really are not suited for close spaces and need help. The

ship with a load is thousands of tons, difficult to move, and hard to stop once moving.

Therefore, they are assisted by tugs in a procedure called 'warping. The cargo ships are tied up with their bows upstream.

They are warped around, not able to make a U-turn. Our tugboat, like others, had a tough hull with a rub rail all around. Auto tires were used as fenders.

A large mat of hemp rope was fitted on the bow. From the wear of many uses, it has an appearance of a walrus mustache.

The ship has all mooring lines in except the spring line. The ships rudder is turned towards the dock.

A line runs from the stern to the tug. The ships engine and the tug swing the laden freighter at 45° away from her berth.

The spring line is trailing. A signal is given, the slackened line is taken onboard. With the slow astern power, the ship backs away from the dock.

Now crosswise in the river, a constant pull on the stern by the tug brings the stern upriver. As the bow points downstream, the river pilot takes over for the 150 miles down to New York Harbor. We recover our lines and go alongside the ship, now under its own power.

The harbor pilot comes down the ladder and boards our boat. He's dropped off at the empty berth and we proceed home

SELF-EMPLOYED
An American tale

My father was a teenager during World War One. He and two brothers left their home in upstate New York and traveled to the big city.

The influenza pandemic was causing the war to stop. A state of confusion reigned everywhere. News stories contradicted each other, and leaders spoke without saying anything decisive.

So, their dreams of enlisting and glory came at the worst time. They each found work in different fields of employment.

Mike the youngest became an apprentice to an electrician. His main duties were pulling wires to the pipes leading to gas lamps which were falling into disuse.

Andrew, the oldest brother, had previously worked on the railroads and was a
of the road was undeniable. As popularized by the *Keystone Cops* and *Laurel and Hardy* – autos were fun

Looking back at the Ford Model-T – it was a primitive mechanic on steam-powered road building machines such as steam shovels, rollers, and such.

Tony, my father, was trained under a Navy sponsored program in the newly introduced technique of electric arc welding.

At the shipyards, ships were being riveted since Civil War days. Arc welding was being introduced everywhere.

They were hard workers and sensible with money. Rooming together, cooking at home, and staying out of taverns – they were doing good.

The roaring 20s, to this trio, meant the sound of the motorcars. Soon all three were involved in a romance that engulfed America. Everyone was in love with automobiles – even songwriters.

The first autos they got to work on were imported to New York in subassemblies needing to be assembled and made roadworthy. It was inevitable that America would have its own home-built autos. Clearly there appeared a variety of machines as there was no standard design. The competition was fierce. The three brothers were also opinionated about autos.

Andrew favorite steam cars. Stream was an accepted prime mover. Steamboat, railroads, and heavy equipment were widely used. Parts and labor were readily accessible.

Mike believed the electric auto have already proven to be the winner. The elegant and very quiet auto and the dray carts on the docks were here to stay.

But Tony argued for the gas-powered machines. The ratio of horsepower to weight had proven they were used in aircraft. There were races attracting crowds to the ranks of auto fans.

In rebuttal, each mode of power had its drawbacks. The steam cars had an insatiable thirst for water that limited their range. Also, when starting from cold, time-consuming waits were necessary.

Electric cars paid a penalty for the heavy batteries which needed recharging and servicing.

Gas buggies had many more moving parts, which needed attention, but fuel was easy to obtain. Most early autos were super heavy town cars – expensive and seldom seen outside of cities. Then, along came Henry Ford.

Ford did not invent the automobile. Just like Westinghouse did not invent the air brake. He improved hem. Edison improved the lightbulb, and the Writes made improvements to aircraft.

Fords success was offering a simple, low-cost machine.

The three brothers were eager to become involved in the automobile revolution. They soon had a garage in Yonkers. Being excellent mechanics, they became excellent salesman and instructors. You see, in those days you had to teach driving to people who never had even touched a car. Also, trade-ins were the source of stories. Auto financing was in its infancy.

The lore design. They were much concerned about the manufacturing costs of labor and materials. In the interest of cost savings, there were no shock absorbers, electric systems, cabin heat, adjustable seats, etc. The oil-lube system included the motor clutch and transmission. Operating the 'T was mostly a physical routine of feet arms and strong back. But everyone wanted to get on the road, at any cost. Business for the three boys was good

Millions of Ford model Ts were produced over many years. All exactly alike. Other car producers introduced cars with advanced improvements. So, Ford was forced to change.

In meetings with dealers, announcements were made to change from the Model T.

From a clean sheet a new design called the Model-A was born. Dealers, mechanics and all involved were stunned. Many spoke of doom for Ford.

"We have for years taught America to drive with their feet," was a common remark. The model-A1, when introduced, proved to be the next generation.

Much improved but was need for further development. I learned to drive on a 1-ton Ford truck called a model AA. (That's right. AA.)

I recalled my experiences elsewhere.

The three brothers, older and wiser, moved back upstate, married, had families, and satisfying lives. They delighted in retelling their adventurous.

CHARLIE

Charlie was an old fella that didn't have a career. He just had a bunch of jobs in a lot of different places. People he met did not care about his stories, and his family and friends heard them all several times.

Well, he was going to go out and have some fun, at that new ski place that opened on the mountain. So, he took the money hidden in the toilet flush tank and set out. But he had to go back in and get his teeth.

He got into the great room of the lodge and looked about. He sat himself on a barstool at the bar and I looked at all the girls in stretch pants and bulk knit sweaters.

The glass wall at the end of the A-frame viewed the slopes and ski lifts beyond. He specifically liked a heated swimming pool, which was both indoor and out. Some extra-pretty, young bathers were making laughter as they splashed.

The barkeep interrupted his staring, and he ordered a drink. He flashed his role of money and, almost immediately, a female moved close. She was a 40plus, old hen, not a chick. But she knew how to use her wiles. They went up to her room, in the adjacent hotel.

Charlie very carefully folded his pants neatly other chair. Placing his socks in shoes set together. He remembered to do this so it wouldn't appear as though he acted in haste went into the bathroom and when he returned, she was gone.

Complaining did no good. He was advised to leave before he was charged with several counts. Back at the bar, he ordered a drink but couldn't pay for it. The bartender summoned a bouncer. In a minor scuffle, Charlie fell off the stool and hit his head drawing a drop of blood from a small cut.

A few of the staff appeared and try to keep things quiet and not draw attention. Charlie was taken to staff break room. Bandaged and calmed down, he was consoled. Instead of getting any bad publicity, an assistant manager decided to get rid of him.

Admonished to keep quiet, he was fitted with rental skis and poles and issued a tow ticket. He was escorted to the ski lift and sent away.

The ride up the mountain in the lift was refreshing – the air cool and serene. He did as others were doing.

The slopes and trails in ski resorts are arranged according to the expertise of skiers. Beginners, such as Charlie, could use the beginner slope. Trails are designed designated novice, intermediate an expert. And they mean it. Another consideration, the skier is under the care and protection of the resort, while on the lift. Should a person choose to not ride a lift back down that's their own concern.

The slopes and trails belong to landowners, usually farmers and leased to the resort. During the non-skiing season of the year the meadows are grazed, and the forests lumbered.

When it comes to these multimillion-dollar enterprises such as a ski resort, things are tidied up legally. Investors use law firms to protect their interests. To moderate risk, the enterprise is divided.

Charlie only wanted his 'Chicago roll of money back and to forget the whole thing. Especially that when coming down the mountain at breakneck speed a tree jumped in front of him! Everyone knew he wasn't an expert skier.

Avoiding all the legal doings, the hotel, the lodge and the lift and the slopes are all incorporated separately. Lawyers represent each separately. Our friend was advised to seek council. What does seek council mean?

The hamlet (not known) where Charlie lived had only one lawyer. He was the county representative and held office in the barbershop where he cut hair. Doing law work against those big city boys scared him. But one cannot ignore the many maneuvers which demanded responses. Months past into years and finally a settlement was reached.

The result was that the plot of acreage with the tree on it was now Charly's. The farmer's Homestead could not be touched. His medical bills and lawyer fees took all the settlement. There was a nondisclosure clause in the signed documents. This meant, Charlie was the tell no one how he got screwed. He had intended to tell the world. He might be living better in state prison, with room and board.

Instead, he tells anyone who will listen how he came to own a tree in Vermont.

CARGO HANDLING

I worked on the docks while still in high school, loading and unloading ocean freighters.

Most people use the words stevedore and longshoreman interchangeably. The facts are longshoreman are dockworkers and stevedores work on the ships.

The word stevedore is from Greek meaning 'to stow.'

Responsibility for the ship's hull and its cargo rests on the ships chief mate. He prepares a cargo plan, which is rather complicated. The many considerations are interesting.

The trim of the vessel refers to the balancing, left and right. That is port or starboard. Trim also refers keeping the ship level by the bow or stern.

'Storage factor' is a term meaning density. Such things as pig iron, copper ingots, steel plates and shapes, are heavy and take up little room.

Many items are less dense and use up volume of space such as automobiles. Specific cargoes cannot be over-stowed - such as autos.

General cargoes and crates or barrels are packed tightly, so to limit movement in the open seas. Bulk cargoes, such as grains, or loose materials as cement are usually put in specialized vessel.

The cargo plan needs revisions as a voyage continues and several ports are visited. Over stowed cargo is to be avoided. Taking out some cargo, which is beneath other items can be costly.

Some ships have cargo spaces that are specially built for multi-use. Refrigerator spaces for an example. Also, spaces called tanks have gasketed covers and are used for extended-range voyages, when filled with fuel.

Devices called 'Butter-wall' cleaners are used to scour the spaces used to carry liquid cargoes such as wine, cooking oil, even lipstick and cold cream.

Reading a ship's manifest can prove interesting. From jet planes and sulfur acid as deck cargo, to guano (bat droppings) and specie (gold coins)

The chief mate really earns his wages. If he does well, he may get a bonus.

A YOUNG FARMHAND

During World War II most able-bodied young men were either drafted or enlisted. Thus, making a severe shortage of hands to work on the farms.

So, schoolboys and older retirees went to the farms, specially doing haying. In upstate New York, the common hay grown was 'Timothy.

It must be cut, 'tedded and put away in a precise manner to preserve its food value. If put up damp, spontaneous combustion could cause a barn to burn.

Oats and other grains are harvested for the seeds. The stalks are gathered as straw and used as bedding. Also, silos are filled with chopped corn, preserved for winter feed for the cattle. Horses were the common power for these operations. Tractors and bailing hay were yet to come. There was a great need for farm help during late summer.

The so-called 'Milk Shed in New York, consisted of the Hudson and Mohawk Valleys, and Lake Champlain areas. Herds of cows on farms provided milk for the eight million in the city. Milk was gathered daily and sent on high-speed milk trains to creameries in the New York markets.

My experiences, like many other boys, school vacations were not vacations. The two summers of 1944 at 1945 will live in my memory forever. Big and strong for my age I was sent to a farm in Saratoga County.

I was to help out on a farm owned by an old Dutch couple whose two sons had enlisted, one in the Navy and the other in the Marines. I didn't know it at the time, but all the equipment and tools were shot. Broken old junk. Keeping the farm operations going was a 16-hour day.

For instance, the antique Ford tractor was hand-cranked; large steel fenders covered the spike wheels. Toolboxes built into the fenders were in constant use.

Because of the spiked wheels, this machine was not to run on unpaved roads. The seat was a formed steel sheet. Don't forget the cast-iron radiator tank with a large oval filler spout, suited for pouring water from a bucket.

I was getting a course in mechanics. The flywheel engine was of great interest with a 'make and break ignition. Living near the Hudson River, I remember the distinct sound of these common motors powering pumps on the wooden barges, passing in the night.

The horizontal single-cylinder engine had a system of firing one or two air-fuel charges, then coasting for several revolutions. Ask any old timer. The cylinder was encased by the open top water jacket. Adding water to the top was a constant concern. This primitive design was in use everywhere, introducing a generation into the world of mechanics. However, the unit as used on this farm was applied in a clever arrangement.

Perhaps the original system for milking cows, to replace hand-milking, allowed farmers the ability to milk more cows, increasing production. The apparatus was vacuum driven.

The electrification of farms had just begun. A source of vacuum to power the milk machinery was devised.

While World War II raged on, with the latest and finest of machines, the home front repaired and improvised. The fly-wheeler engines were in use on farms and elsewhere, to power pumps, saws, grain mills, construction equipment – you name it. It was the forerunner of Brigg-Stratton type of small motors.

As used on this farm, the engine was used only as a source of vacuum for milking. The intake was connected by pipe to an antique copper hot water tank. The tank was fitted with a very delicately sprung disc valve. Thereby, a tank of lower pressure air to power the system.

Piping ran above the necks of cows, with a disconnect fitting. Hoses to which the teat cups and milk container were attached, were moved from stall to stall. The warm milk was fed from milking machine container into large milk cans to be cooled and shipped.

The large, heavy cans, filled with whole milk, well were placed in a spring-fed concrete pool. Sent to creameries where the fluid milk was separated, pasteurized, and made into many dairy products.

Like other enterprises the mom-and-pop dairies have been modernized and incorporated into more efficient, redeveloped, features of our society.

Cows are no longer called by name – they are just numbers. But you can buy a gallon of milk anywhere and drink it down, delicious and healthy, with safety.

A SMALL RAILROAD TOWN

I have a brother-in-law, Gert Larvig, M.D., who lives in Florida. Now retired, he has a hobby of model railroading. As a member of a group of miniature railroaders, he became involved in a rather large project.

The club has undertaken to make a huge layout of HO scale tracks and scenery. Divided into small teams, each team is assigned a section of the overall plan.

Upon a four-by-eight-foot sheet of plywood, the rails are connected when completed and a sophisticated electronic control system allow several trains to be operated at the same time.

One group made a rural scene with farms and livestock. Another built a fishing village, another a lumber mill and pond.

The many panels were connected would be stretch over 100 feet. When viewed in its entirety, one can only marvel at the details, representing much patience and time.

Each year at holiday time, arrangements are made for a public display, usually at a shopping mall.

The assembled display, complete with sound effects, is open up for public viewing. A donation box encouraged folks who enjoyed the scale modeling, to donate to local charity.

After the holidays, the elaborate display is dismantled and plans for next year are formulated at the next club meeting. Perhaps you could become involved in a small that is, miniature way.

TROPICAL STORMS

Tropical storms, such as hurricanes are very wet and windy. They can be destructive and cause major changes in our communities. Those of us that live in the gulf region of America, take heed when severe weather warnings are posted. History is full of accounts regarding destruction, flooding and power outages, as well as casualties.

However, people of Bermuda have a different attitude toward these tropical disturbances. In fact, they may go down to the church and pray for a major storm to come their way. The very existent of the Bermudan society depends on these visitations of wet storms.

You see, there is no ground water source on these islands. Based on coquina a soft material made of shell, there is little soil and no streams or lakes holding water. Unlike most civilized area, there are no water mains in the street to be connected to for service.

Large enterprises, as hotels, civil buildings and military have rather big catchment. That is concrete line swales to catch and store rain. There are businesses that deliver water at a price.

The construction of a home here is unique. Quite different from our Florida houses. All construction must be approved and closely supervised.

The area for the house is swept clean and a specialized method of cutting and storing blocks of coquina is begun. A rectangular pit the shape of the floorplan is dug, much like a swimming pool. The cistern is parged with a waterproof cement.

The blocks harvested doing excavation, are used to build thick perimeter walls. Door and window openings are fitted with heavy lintels.

Support for the roof is rather impressive. The walls and roof are built hurricane-proof. The structure is strong.

Most residences in the southern United States are frame with extensive use of 2x4 lumber. Lightweight trusses sheathed with chipboard panels support the roof. They are covered with tarpaper and shingles.

Building codes require a horizontal wind load of 100 PSF, and a roof vertical load of 40 PSF. Perhaps the greatest load on these roofs is the roofers. Also, the low pitch grooves act like primitive airfoils and lift in high winds.

The Bermuda roof is framed with very heavy four-inch joists and fitted with 2 x 4 purlins. The actual roof coating is a system of concrete slabs.

About a foot wide and two feet long and about two-inches thick they are fitted overlapping and finished with white parched cement. Many have lasted over one hundred years.

Gathering the rainwater is the most important thing. A unique feature is hinged water chute at the bottom of the gutters and downspout. The first amount of rain is allowed to clean the roof of debris and droppings. The cleaning water is sent to the yard. A small box on the shoot, once filled with water, will tip, and send the clear rain to beneath the house.

A small trap door in the kitchen floor, allows people to view the reservoir. Many have a few goldfish as indicators. Should the fish show any problems, a gallon of bleach is added.

Across the great blue marble we call Mother Earth, many people have survived by implementing unique solutions.

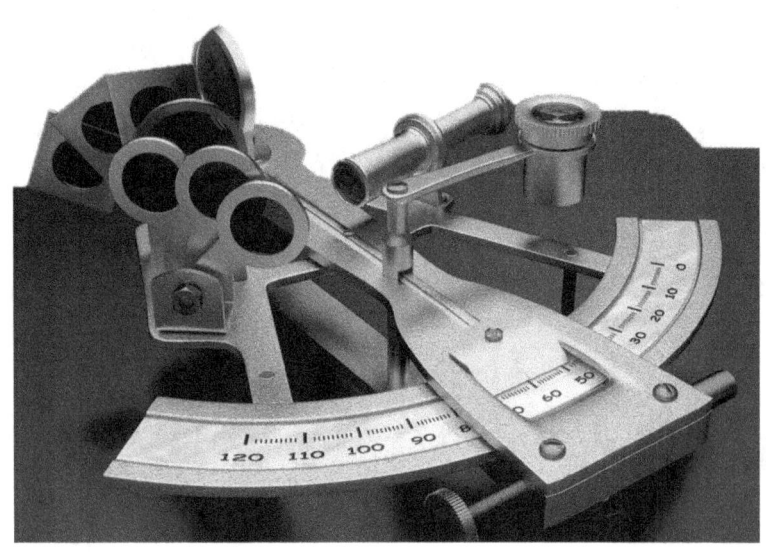

WHAT TIME IS NOON

I went to sea as a young man on a merchant ship – the Buckeye State, of the States Marine Lines. She was a victory ship – steam turbine of 8500 hp. I was assigned to the 12 to 4 watches on the bridge. Usually, eating with the on-watch officers in the saloon. (Merchant mariners don't have ward rooms.) The second mate is the navigator and I'd ask him what time is noon?

You see, the time used by your wristwatch or wall mounted clock depends upon which time zone you're in. This means people in New York, Buffalo and Pittsburgh have the same time settings. This arrangement was set up to accommodate railroad timetables.

Ships at sea calculate a day's run from noon till noon. Local-apparent noon. A position when the sun is directly south of the ship. An accurate sighting

My first vessel, Victory ship Buckeye Express

allows you fix the ships latitude, longitude and miles traveled.

As a trainee, the noonday fix was plotted by sights on handheld sextants. The captain, second and third mates, and I plot and record our information.

Traveling the oceans of the world, the clocks were changed for each 15° of longitude. Timekeeping may take some consideration. For instance, Ferdinand Magellan led an exploration to circumnavigate the earth.

He was killed in the Philippines. But his crew completed their return to Portugal. They kept track of the days and planned to attend mass on Sunday. Only to be told it was Monday. They'd lost a day.

And Phineas Fog traveled eastward in Around the World in Eighty Days. Arriving back in London, he gained a day.

I racked my brain and studied hard to learn the use of the sextant.

Back in the 1950s, techniques were those used during World War II. Electronics coupled with satellites, have changed everything. The use of the sextant was made an option at the academies. I turned in my instrument.

However, after a half century of being ignored new light has revived interest in celestial navigation. The joint Chiefs have considered 'What would happen if electromagnetic forces were to knock out navigation satellites?' Programs to re-introduce the use of the sextant are being considered!

HEROES I HAVE KNOWN

History has preserved memories of great leaders – distinguished military and political men and women. Books, movies, magazines, and radio have recorded for posterity the well-known and highly decorated ones.

But I've worked with and have several heroes who should be recognized as they are and perhaps on this great and there are a million more.

For instance, John Draper a member of the faculty I worked with. He had two ships torpedoed from under him in one day! Before Pearl Harbor and Hitler declared war on the United States. As a merchant marine, like many others, he sailed on a mobile tanker vessel while war raged in Europe. A German U-boat torpedoed his ship, and the crew took to their boats. Their distress signal was answered by a royal navy Corvette, who picked up the survivors.

The U-boat waited and torpedoed the rescue ship. This occurred north of the Azores Islands. Eventually, whalers from the islands picked them up. The crew was sent to Lisbon, Portugal a neutral country. The prefect kept allied survivors on one side of the Tagus River and Germans on the other. One day a week the sides were alternately allowed to visit the center of Lisbon.

After some time, a prisoner swap was arranged, and John and his buddies were flown to neutral Ireland. They walked across the border to Northern Island. He joined the U.S. Navy. "I want to sync a few U-boats," he said.

Another great story – my friend and Co-faculty, was John King. Like many teenagers in the depression years, he grew up in Bristol, Virginia, and Tennessee. The state line went through the middle of town. It was easier to move, then pay rent.

John joined the Marines and distinguished himself as a sharpshooter. He tells of competitions where rifleman shot at a ships bell, about the size of a garbage pail. Each round of shooters backed up. Eventually to nearly 1,000 yards! The Navy needed gunners who could hit their targets.

He was assigned to fleet Marines on a new aircraft carrier. USS Wasp CV7. The ship left without aircraft on board on a secret mission. The gun crews fired familiarization practices from gun platforms without tubs for protection.

They went to England and were boarded with army type 'Hurricane fighters. They were going to the relief of Malta.

Once they pass Gibraltar, there was no secrecy. Once within range of Malta, the planes were launched. These land planes could not use the carriers catapult and were not equipped with tail hooks. So, they could not be recovered. Pilots were told to ditch, in an emergency.

The Wasp return to the states took on air groups and practiced, on the way to the Pacific.

John was vetted and now a gunnery officer. The Wasp was badly damaged, and the hanger deck, which was piped with aviation fuel, was set on fire. This ran the full length of the carrier

Finally, with a pronounced list, the order was to abandon her. John King stepped from his gun mount onto a destroyer and never hit the water.

Given leave, his mother worried about his dress uniform and sword at the bottom of the Pacific. He never replaced them.

Another member of our faculty, Harlan Crawford, has a tale to remember. He grew up in horse country, in Kentucky. Like many young men in the poverty-stricken depression, he joined the army.

He excelled in the cavalry and played on his regimental polo team. As war approached, Harlan was involved when the army mechanized. Horses and mules were excessed. Troops had to learn to drive vehicles.

As a last great March, the entire regiment left. Fort Bragg, and trail-marched to Fort Benning, Georgia and then to Fort Hill in Oklahoma, the 're-mount center. Stable sergeants wept and predicted the army had made a mistake.

Eventually, Harlan was trained as a paratrooper, and as an officer, he led of his own unit.

Paratroopers were elite soldiers, dropped first in an action to secure an area as the vanguard in battle. That is, once on the ground, they were infantry men.

Most paratroopers jump into combat once. But fate and circumstances had Harlan jumping into Sicily, Italy and later into France. On D-day. He was the executive officer of the 192nd airborne. He was a warrior.

As a fellow member of the same faculty, I moonlighted as a designer and made a set of plans for Harlan, for a house to be built in Silver Sands by himself and his sons.

As time went by, I worked at the Kentucky space center along with another combat veteran was Harlan Maye. His exploits were in the so-called Korean War which has not yet ended.

As a young man he and a few buddies enlisted. He was trained a Ranger, improperly called commandos.

He distinguished himself by his daring. Several times he parachuted, behind enemy lines. In darkness, using stealth and silence, he had to cross back across the battlefield to report. This required being him to be at a certain time and place. Usually under an arranged hail of harassment fire, the danger was being killed by friendly fire.

The last individual patrol, he was wounded. Thirty-seven rounds in his pelvic region ended his combat career. After many surgeries, and many months in recovery he was sent back to Fort Riley with no duty assignment. He was given 100% medical discharge into retirement. Given employment in the aerospace programs he lived his final years in Cocoa Beach.

The last of my personally connected heroes, is my cousin Andrew. Nicknamed 'Captain, he was a young man unable to find work in the 1930s. He enlisted in the Marines after being in the Civilian Conservation Corps. He was a sergeant and stationed in Guantánamo, Cuba when war broke out.

When the Japs bombed Pearl Harbor, a high alert was sounded. Everyone knew they would attack the Panama Canal next.

In rapid fashion, they were deployed and set up defenses for the vital canal.

But it didn't happen. Instead, the Japs went South to Borneo, where the great Dutch oil facilities were. So, the long Pacific war began, the business of island hopping.

My cousin Andrew landed from landing crafts, under enemy fire, with many comrades. Our families prayers and our novenas must have protected him. Through Hell and high water, he went all the way from one campaign to the next.

Except in the last great battle, Okinawa, he contracted jungle rot in his feet and was hospitalized. He was mustered out with many medals and a set of crutches.

Returning to Troy, New York he married had three sons and a daughter. Working as a Carpenter, he doesn't speak of his exploits. He became a bitter old man.

Once he told my brother and I, the following. "If I saw a truckload of bananas, and that bananas belong to the government, even if I didn't like bananas, I'd steal them."

For my own story, I volunteered three times for overseas. But I never got out of the training command. Placed in special services my unit did everything no one else did. From registered couriers to prisoner catching, medical librarians, to base education – you name it. All on unappropriated funds

These heroes I knew 'Saw the Elephant'* and should be remembered. I take a little pride in calling to mind Winston Churchill saying, "They also serve, who only stand and wait."

*To get real-world experience at a considerable personal cost. (A popular saying in the mid 1900s)

PLIMSOLL MARKS

Did you ever wonder what those marks on the side of a ship stand for? They are called Plimsoll Marks indicating the depth to which ship can legally be loaded.

In ancient times a brass nail was driven into a ship, indicating its registered tonnage. Modern freighters have several additional markings, for instance the mark for freshwater indicates loading at New Orleans or the Great Lakes as the ship will ride higher than in the saltwater seas. Because freshwater weighs 62.5 pounds per cubic foot.

The Indian Ocean is the saltiest of the oceans. Also, seasonal hazards are also considered. The north Atlantic in winter is considered to be the most hazardous. Therefore, ships carry less tonnage and have reserve buoyancy.

In January, crossing the north Atlantic, we were loaded down to her marks maximum. I started the bridge watch, seeing heavy seas come over our bow, throwing tons of water on our deck. As the ship rolled and struggle to rise, watching the seas pass around the hatches gave me pause to repeat a common remark. " I hope this ship wasn't built on a Monday."

Perhaps it is more proper to remember the song, that goes something like this, "They that go down to sea in ships, they see the wonders of the Lord and marvel at his work."

The song quotation "Asleep in the deep," cautions "Sailor beware, beware…"

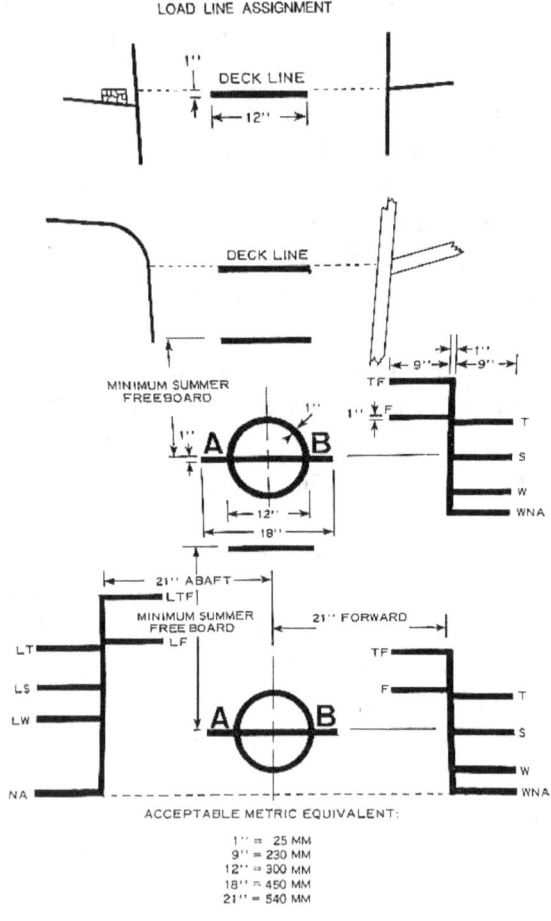

Fig. 5 Load line markings on a ship's side

A TALE OF MONKEY FISTS

A friend of mine named Frank worked on construction dredges for many years, keeping the harbor and channels usable. He was hit on the head by a swinging shackle. He recovered with a plate and was pensioned off.

We both were volunteers for many charity projects as brothers in the Knights of Columbus. As former merchant marine seamen we certainly knew our knots.

The monkey fist is not used in many applications but is commonly found at the end of a heaving line. Usually the size of a baseball, at the end of 100 feet of line a little thicker than clothesline.

The monkey fist is tied about a big rivet or nut, for weight. If used, a crewman will heave the line from the ship to the wharf. Do not catch the ball. It will break your hand. Grab the messenger line and haul the hawser to a mooring bit.

Now, as a hobby, Frank tied many smaller unweighted monkey fists of decorative cord. Much like croqueting or knitting.

He would while his time making ten knots, evenly spaced. One for each decade of the rosary. Adding other knots in a loop, with a cross at the bottom. It was hung for display on two hooks. It was bigger than a window space and much admired.

Over the years several dozen were giving to churches, schools and hospices.

As a joke, I made up a story wherein Frank went too far. The neighborhood cops stopped him, when taking bowling balls from the bowling alley. He took Frank to the station house. When the Desk Sergeant asked what he was taking those bowling balls for?

Franks reply, "I am making a rosary for the statue of liberty."

BIRDS OF A FEATHER

I've been fascinated with all things to do with aviation. From building model airplanes as a boy, being a member of the Civil Air Patrol Service in high school and serving in the Air Force.

Over the years I held a commercial pilots license with an instrument rating. It was hard during the 1950s – trying to keep current, earning a living not having a pilot's job.

The best I could do was third pilot, on flights lasting over eight hours, in a C-47.

Because of my apprenticeship at Watervliet Arsenal, I got a job as a draftsman. I had a security clearance and was sent to Lincoln Laboratory, ONR,* helping on the BMEWS** and DEW*** defense projects.

My brother, brother-in-law and a few friends stuck it out and landed flying jobs. I particularly remember Vinny Renard.

He was a sharp pilot and wound up as a captain with Eastern Airlines. Airline pilots were scheduled to fly 80 hours a month.

Vern would fly from Miami to Mexico City and back, then Miami to San Juan and return. He lived and breathed flying and was good at it. One of his close friends was Martin Caiden.

*Office of Naval Research
**Ballistic Missile Early Warning System
***Distant Early Warning

Martin Caiden was related to Bill Piper, aircraft builder. Marty became known as an author, screenwriter, and airshow personality.

He owns a WWII German JU-52. He kept the three-engine aircraft at the Merritt Island airport next to where I lived. They called the taildragger 'Iron Annie.

Ever the showman and teller of tales, Martin often displayed the ship at airshows where he was an announcer.

I worked for Associated Engineers of Massachusetts at Lincoln Labs, Inc. in Florida. So, I was vetted to Orlando. The Cold War was "hot."

Those involved were serious and dedicated. Weapons development projects were very serious. The missile plant we activated in Orlando it was operated by Martin aircraft.

The weapon systems include Lacrosse, Bull Pup, White Lands, Maze, Matador and Pershing. Scientific advances were introduced at a terrific pace – especially electronics. We rapidly changed from vacuum tubes to printed circuit boards, miniaturization and solid-state. Then, the race to the moon began.

$93 billion was spent doing the Mercury, Gemini, and Apollo programs. At its peak, 23,000 people worked at the space center. I had become a supervisor over second-shift documentation in the headquarters building.

One of CAD draftsman working there was Cliff Rupp. Born in Brooklyn, he was an avid baseball fan and player. He was chosen for a tryout in major-league farm team but didn't make it.

Cliff and a partner went into the road sign business. They had a truck with a boom and leased a warehouse for storage and making billboards. They were doing well, until Lady Bird Johnson's highway beautification program put them out of business. Being resourceful, they went into the worm business in the leasing building.

The worm business is labor-intensive. But can be profitable. Not only worms, but their castings are sold to florist nurseries.

One night, a tremendous thunderstorm raged. The next morning no worms were found in their beds. They found their advisor and expert. He told them that the worms have stampeded.

"What to do?" – "Go get them back."

The area was surrounded by buildings on concrete slabs. So ended the worm business. But the superrich soil in the beds was to be sold.

Cliff and his wife lived in one of the high-rise condos on Cocoa Beach.

After work each night, he would go up in the elevator. The elevator car had an advertisement for Eastern Airlines posted on one wall. The ad campaign consisted of a picture of a veteran pilot and a captain noting safety and service. One night, he got in the lift with a uniformed pilot. The pilot and the picture were the same. The name on the ad was Vern Renard.

As we worked together for months, Cliff knew I had an interest in aviation. I guess I mentioned flying at Glens Falls or Vons name. That connection was made.

I phoned him and we got reunited. I asked him, "How, come you're in Cocoa Beach?"

He told me he was a longtime friend of Martin Keaton. He flew 'Iron Annie.' Insurance restricted who could pilot the Junker. Martin called Kayden who did not fly it but took many photos.

A true turn of fate caused me to be involved in a totally different realm. Because I had a lot of experience in aerospace and a Rank-two (Masters) teaching license I was contacted by a Dr. Dave Stewart.

He had been granted funds to do some curriculum development. I was offered a small stipend to participate. The project was to develop aerospace education in our Vocational-Ed programs to satisfy the needs of our growing space community. In order to accomplish this, we needed to survey the field, develop goals, and make presentations,

The best part of this undertaking was Dr. Stewart had a Cessna Skylane airplane. I've had many hours in 1972s an all-metal, four seat, high winged craft set on tricycle landing gear. It was a popular cross-country ship.

Dave Stewart and I contacted many people with interests in aerospace. Of course, we flew to interview owners and operators throughout our region. We tried to have all parties put in their two-cents worth.

We took notes and recorded inputs of many air service professionals. A most interesting person was Dr. Joseph Hunt, president of Embry Riddle at Daytona Beach.

In 1958, when I got my instrument rating at Embry Riddle in Miami, we flew training flights from Sweetwater field.

Joe Hunt was fascinating for many reasons. He has had more hours flying airships and anyone in the world including any of the Zeppelin airmen. From navy free balloons, blimps on anti-submarine warfare to research and clandestine long-range flights. A flight in a lighter-than-air ship is a rememberable thrill.

Just a note: Bud Oliver, operator of Volusia Air Service, kept a Sikorsky flying boat, made of stainless steel. That ship was older than me and in better shape.

Well…time and circumstances change, friends pass, and we lose touch. Sadly, Vern Renard lost his life in an accident at Parker Dunn airfield

Around the space center there exists a great number of aviators. From restorers to homebuilt and kit planes. A trip to local airfield is a delight and hanger talk is always interesting.

To paraphrase the river rat, from Wind in the Willows, "There's nothing better in life than messing around in airplanes."

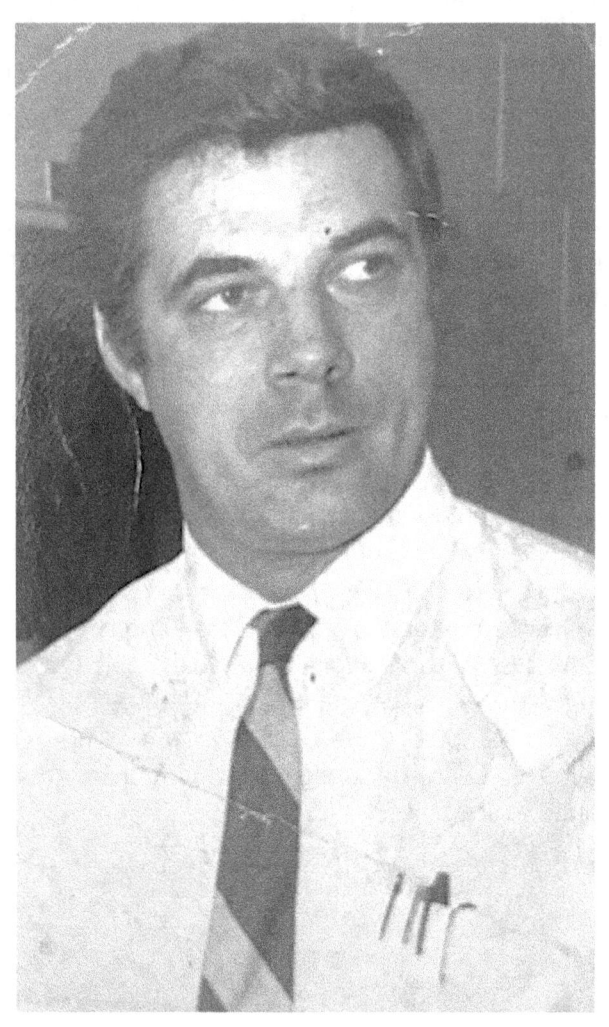

FIRE ALARM SYSTEMS

The town I grew up in was cut up by its many ravines. The streets paralleling the river were generally straight and level. Moving away from the river, the land rose up and was home to many streams, creating a problem.

Neighborhoods were built up, making streets segmented. When locating an address, one must approach the correct portion of street. In steep slopes roads were separated into upper and lower single lanes. Very steep downhill streets were not paved and not used during icy conditions. These matters caused local authorities problems – firefighting for instance. Most of the town was comprised of wood framed homes, with common walls.

Being a rail center constructed in the mid-1800s, a popular arrangement evolved.

Fire houses and other horse drawn equipment, were privately owned, and referred to by name. We lived near the Mink-Claxton fire company. To this day the fire trucks do not have numbers on them.

Before the 20th century, a crude fire alarm system was made. It consisted of a tire off the drive wheel of a steam locomotive, suspended by a chain under gallows made of rail ties. A steel rod was used to arouse the neighborhood.

Fire companies only responded to houses that paid for protection and displayed a proper sign.

Electricity was introduced and homes abandoned oil lamps, but many still used fireplaces and cast-iron stoves. The alarm system was made of pole boxes located every several blocks, mounted on telephone poles at intersections. Each pole box had a number. An air mounted loudspeaker atop each fire house blasted the number of the alarm box, thereby announcing to the entire community its location. Volunteers were to go to the alarm box location.

Whoever pulled the alarm should stay put with the information. The fire truck would arrive, and the driver commandeered the use of equipment and crews.

Normally one alarm brought the nearest company in response. If necessary, a second alarm brought all city forces and three alarms involved mutual aid. Fire alarms brought a dynamite specialist.

Fire brigade turned out in uniforms, marching with their bands on special occasions. Teams competed in firefighting contests to win awards.

Until After World War II, half the homes had no phones. With the proliferation of phones and the introduction of 911 services the red cast iron boxes on poles have disappeared. The hard paper cards listing the pole boxes are collectibles no longer offered by businesses.

The practice of neighbors jumping out of bed in the middle of the night is gone. Also, perhaps, the pioneer spirit that motivated volunteers.

ICEBOATING

One day in January, my brother and I wanted to go to Saratoga Lake at the invitation of his friend, to experience iceboating. It was a bright sunny day, with a fresh breeze. Both my brother and I were experienced sail-boaters but were anxious for a new venture.

Iceboats have been around for a long time. Mostly one-off designs of varying sizes and shapes. I recall seeing a picture of Roosevelts' exceptionally large boat, of a century ago.

Since the 1930s, in order to promote iceboating, two most common designs prevailed. The Yankee

iceboats and DNs. I was told DN stood for Detroit News who made the plans available to homebuilders.

There were a half dozen boats on the lake that day. Similar in design, each boat carried a tall narrow sail, with full length battens and many parts to the main sheet. There were three steel runners under the cruciform wooden frame. Each craft could carry two people.

Both of us took rides, with the owners and had many grins and noticed similar comparisons. For one, sailboat really go much faster than the wind, even on a reach. Another is weekend sail boaters are told never to jibe. That is do not have the wind cross your stern. "When putting about, put your bow into the wind."

Iceboaters ignore these axioms. In a breeze of fresh wind, these boats are traveling twice as fast as the wind. The wind never catches up. Single-handed they could hit 50 miles an hour on the frozen lake surface. There is no wave action. Only the noise of the skate like runners sliding across the glass-smooth surface.

Be sure you have plenty of room to make wide-sweeping turns. These boats all also have no center boards or keels. So, the depth of water is no consideration.

Stopping requires some thought. Coming into the wind, the sails will luff, but the boat may back up!

The sun was low in the west. And a chill hit us. We thanked our new friends and invited them to sail with us on some warm summer day.

INTERNATIONAL CODE OF SIGNALS

When the radio communications were first used on ships, it was only CW signal – that is a series of dots and dashes called Morse code. The navies and shipping companies sent messages by code. They did not have so much as an SOS signal. Soon telegraphers were earning money, sending messages for passengers. They each had their own system. For instance, birthdays, Christmas, and wedding correspondence would be numbered and typed out fully when sent to the recipient. But disaster struck, ending these selfish motives.

During World War I, ships would rendezvous in Halifax, Nova Scotia to form convoys bound to cross the North Atlantic.

Halifax was excellent choice. It had a large harbor with space for anchoring many ships. The inner harbor was entered through an narrow passage from the sea, making it easy to defend and weatherwise.

The war was raging in Europe and vital supplies were amassed to sail. Two ships, going in opposite directions collided in the narrow entrance channel. One was loaded with relief supplies the other loaded with explosive ammunition. The resultant blast caused destruction of the city and loss of thousands dead or injured.

The two ships sunk in the channel prevented traffic movement in either direction. This could alter the course of the war. A marine board investigation found the cause to be cross signaling, meaning no accepted practice existed

The conference included practically every maritime nation. It produced many actions adopted by all signatories, including rules of the road, signaling, safety equipment and drills to be conducted.

I specifically liked the introduction of the Q codes. This book of signals could be used by blinker lights, signal flag hoist, semaphore, or radio telegraphy.

The beauty of the system was that it transcends languages. That is an American sends QRM requesting medical assistant. A French or Japanese ship reads in their own language and responds. I recall, when approaching the Panama Canal, the shore station would request the name of your ship and where you're bound from. Because the signal men are amazingly fast, I replied QRS, please send slower.

The four letters KXFW completely identified our vessel. It should be mentioned that doing the great war both sides abided with the Safety of Life at Sea (SOLAS) conventions. A testament to international brotherhood of seamen.

THE FOURTH ESTATE

There was a mention of the fourth estate on TV and my grandson asked me what that meant. So, I asked him if he wanted a long or short story? He, knowing how verbose I could be, said, "A short one."

To which I replied, "I don't have any short ones."

He exhaled and settled into the sofa with a thump.

The fourth estate refers to the news people, in recent history – now power brokers.

You see, or many centuries, power was kept in balance by three estates. These heads of state (Royalty or Heraldry) as one branch of the triangle. The military has a second side of the arrangement and the third represented the church.

When properly used, this set up a self-righting method.

If the military attempted a coup, the rulers and church would intervene. Should the church assume too much influence, the military would side with the administration. A despotic or reckless king would be taken to task by the others.

Acceptance of the members of these three branches of power came with rules to be followed, cased in traditions, treaties, protocols and what have you.

Then there entered the press, a group who wanted to play self-proclaimed distributors of information (Or misinformation). They began to influence public opinion, and, over time, held great power.

The traditional three estates sought their help and made deals to gain their assistance for pet projects. Leaders, needing the trust of the people, began using various agencies to gain influence and power.

Groups or individuals with money could promote themselves. Not buy outright lies, but by omissions, distraction, innuendos, and a variety of mob-psychology methods

Years ago, prizes were awarded to investigative reporters who sought out corruption in crime. I remember one could walk the corridors of public buildings and talk to almost anyone, search records and report about public business

Today, reporters are badged and sent to newsrooms. There, with chair, desk and coffee, they cannot roam about, but are sent news releases. Thus, the flow of use is well regulated. Investigative reporters on local broadcasts only search out small businesses. They are told to not go after large corporation or government entities, which could be expensive

What this amounts to, is well-managed news – or propaganda. The networks call the news programs shows. Presentation is made by people who are personalities not journalist. Not elected or vetted by a public agency. So-called 'news persons wouldn't dare be creative or mention anything not approved by the invisible higher-ups – the 'Kingmakers.

Of course, you can follow Will Rogers who, long age, said, "All I know is what I read in the newspapers!"

HOG ISLANDERS

Hog Island refers to the marshy area near Philadelphia where many ships were built. The classic freighter was a steel hulled steamer, with raised poop and bow with house amidship.

During World War I they were riveted and during World War II, wielded. As a note, workers were provided with a quick lunch of a sandwich which became commonly called a hoagie.

Thousands upon thousands of rivets were used by a team who had the specific skills and coordination of a football team. The oldest of the group of four stood at the forge. Using coke as a fuel, a few rivets were heated. in the forced air fire.

Care must be taken for steel can be overheated and ruined. Rivets not heated to an orange-yellow heat are not used.

The rivets are picked up with tongs and tossed through the air to be caught by the second team member in a metal, cone-shaped bucket. Quickly the red-hot rivet is inserted into the hole in the plates. A man with a bucking bar places his tool on the head of the glowing rivet.

The riveting gun is held against the red-hot shank to form a head on the hot rivets. When properly formed the rivet cools and shrinks, tightening the plates together in a waterproof seam.

Actually, as the ship sails in the seas, its hull flexes and small leaks occur. Doing annual overhauls, a caulking iron and an air driven hammer is used to tighten suspect seams. The working of a ship's hull is expected.

During World War II a few ships were lost because of their rigidity. Modifications were made to have their hulls less stiff.

REVERSE MORTGAGES
(Good or bad?)

The dream of owning one's home and having a family place in the world is perhaps a concept that is disappearing from the American scene.

Mortgage bankers, politicians and social leaders have combined to change our society dramatically. Several forces are at work in our communities, allowing modern concepts to take place.

For instance, the government sponsored housing project of the last generations, have proved disastrous. Many tall apartment buildings in cities across the globe have been demolished. They proved to be impractical. Concentrating low income and welfare families, also concentrated crime and disorder. Police patrolled in pairs, and swat teams often dealt with hotbeds of dope and alcohol abuse. Single parent families and troublesome schools are all part of the 'projects.

My observations are gathered from having worked at the State Architects Office.

Project buildings, located in central city sites, are very expensive. Steel framed, brick clad, high-rise apartments, require elevators, and strict building codes. Windows, doors, railings etc. are of institutional grade with heavy hardware. Pavement is well grained; lighting is provided to meet commercial standards. In other words, these vertical slums are well-built and expensive to maintain.

Consider the taxpayer, worker living in the suburbs, who pay for these inner-city houses. His home is a lightweight frame or block house, sheathed in chipboard with a shingled roof. The systems of plumbing, electrical and heating are cost-effective units – not commercial grade. They require constant out-of-his-pocket repair. He must also enjoy mowing grass, doing landscaping, and avoiding citations.

A friend works for a company that installs elevators. Building four stories or higher, require elevators. He has repeatedly responded to projects when the car won't go to the basement laundry. Trash thrown in the elevator shaft prevented that.

He also noticed plumbing fixtures and drain gratings removed and sold for junk. All the while worrying about his truck parked nearby. Every night the sirens scream.

So, the do-gooders have come up with a solution. Reverse mortgages are being touted by celebrities across social media. Dont leave your paid off house to your heirs.

Go on a cruise; live it up. Have fun on your retirement. Eventually HUD will take over your house and the inner-city folks will move in, checkerboarding the upscale neighborhood. Our leaders are 'looking out for us.

THE DRAFTMAN

A career as a draftsman is misunderstood by the public. As assistants to architects and engineers, they are variously employed in many fields. An oversimplification may be that they are communicators.

By using images, symbols, and notations, scientific and technical, instructions are communicated to many disciplines, using the vernacular that is appropriate.

When preparing drawings, for electronics, carpenters, steel workers or others, the correct trades must be considered.

Among the many specialized groups of draftsmen, there are common characteristic.

Here are a few clarifications. Mechanical drawing does not mean drawing machine parts or assemblies. Mechanical drawing, or engineering drawing means, 'Not freehand. Line work is drawn with the use of instruments. Guided by straight edges, triangles, compasses etc. Drawings are executed in a classic manner. Properly done, one should not see any individual's influence. Even the lettering should be done in accepted standards.

Sketches or layouts are preliminary working pencil drawings. They seldom leave the drafting room. Permanent drawings in ink are used in making prints. Approved permanent drawings are never folded but laid flat or rolled. Prints are cheap to duplicate and can be folded, filed, mailed, and marked up in use. They are easily replaced

Since World War II, standard sizes of drawings have been universally adapted. Based a letter size of 8 ½" x 11," B size is 11" x 17." C size is 17" x 22" etc. This allows prints to be folded into letter size and stored in filing cabinets.

Historically, the more ancient form of drawings was architectural. By convention, views of building elevations, sections and details are shown with 'up towards the top of a sheet and down to the lowermost edge.

With the advent of the industrial revolution, engineering drawings differ having main views and sections shown in orthographic conventions. There is no 'top or 'bottom to a machine part.

An architect preparing a house for a client will usually abide with local conventions and ordinances to have his plans approved and executed. Large projects are referred to architectural engineering.

A full set of plans may be included civil engineers for site preparation and foundation. Structural engineers electrical, plumbing HVAC and landscape sheets. They will put together a contract

set of many sheets should look as if they were all done to the same standards.

The title of engineer is reserved for graduates of an approved school. Non-graduates, through experienced, are called designers.

A curious aspect of the drafting trades is that no drawing is made unless necessary. Therefore, draftsmen are always making plans of items which do not exist. By conventions and practices, team professionals and their assistants planned and have built marvels of the world we live in.

AIRCRAFT I HAVE KNOWN

My interest in things aviation began as a schoolboy. My brothers and I were fascinated with the reports of exploits of pilots during World War II. The aircraft and those who flew them captured the imagination of everyone.

The war ended when I was 12-years old but, I was hooked, even to this day. We bicycled to a nearby airfield to watch and listen to the sounds of light planes on a grass field. Our gang built kit planes (rubber band powered), at first. By high school, we had powered free-flight and control-line airplanes. Some were home built original designs.

Then, magic happened. Five of us, joined the Civil Air Patrol. As a group we walked across the river and up State Street, each week to meetings. Learning about aviation and going to the county airport was great fun. We even got to ride in light planes, owned by senior members of that CAP.

The first time ever I had my hands on the controls of an airplane, it was an Ercoup. This two-place, four-cylinder aircraft was unique in many ways. That Ercoup had tricycle landing gear in an era of taildraggers. It sported twin rudders and was considered spin proof and had no rudder petals!

In the late 1940s, light planes were pretty much designed of bamboo and bedsheets. (That is, fabric covered, tubular and wood frame.) With wooden props and no electrical systems. Pipers, Taylorcraft, Aeroncas and others were hand cranked to be started. "Give me a prop," was commonly accepted. Only flown in (VFR) visual flight rules, traffic control was by handheld signal lights, where available.

I soloed, and logged hours in an Aeronca Champ. I still love that aircraft! It was head and shoulders above the Piper Cub. Having an 85-horse engine, metal prop, closed cowling, and an auto-type door, it also had a Olro-strut landing gear. That made for improved landings.

Modern design showed up everywhere and along came many improvements in the ships and their equipment.

The Cessna-172 took over the light plane market. As an all metal, high wing ship on tricycle landing gear, the four-seaters had electronics. An excellent cross-country plane, with the use of radio communications and navigation, it really introduced business flying.

I earned my first bucks flying a 172 for Cape Aircraft. Anxious to build up my hours of variety of aircraft including Stinson Voyager, Ryan ST, 40-horse Taylor Cub, Navion and anything else. The fever was still there

Itchy feet caught me going to sea, to 'see the world, in the Merchant Marines. (I'll-treat that topic elsewhere.) Just let me make one comment. You do not join the Merchant Marine! This is not a military service, but private business like the railroads or trucking lines.

I broke articles, went home, and was immediately drafted. My number had already gone by, so I reported as directed. I expected to go into the army for two years but was selected for pilot training in the Air Force.

After completing preflight at Lackland Air Force Base in Texas, I was sent to Stallins Airbase in Kinston, North Carolina. The base had only 600 people and was operated by Servair Corporation.

Most people don't know that the Air Force doesn't train people to fly. Civilian instructors are assigned four students each. Flight and ground instructors are contract personnel and abide by the FAA rules. This group of aviation cadets contained many Mutual Defense Aviation Program trainees.

The aircraft to be used in this program was unusual.

The first ten hours each cadet received was in a PA-18. They were Piper Super Cubs modified slightly for military use. Bucket seats, radios and extra window areas were installed. Immediately upon soloing, we were introduced to the North American AT-six Texan.

This aircraft was the advanced trainer doing World War II. Pilot trainees were sent from AT-6s to combat training.

This was an is a whole lot of aircraft to use immediately after getting a student pilots license. An all-metal, 6,000-pound plane with a 650-HP radial nine-cylinder engine, retractable landing gear automatic propeller and more.

This airplane was very rugged and used as a transition to frontline fighters. It was said, "If you flow T-6, anything else would be easier." There were over 10,000 Texans built. Many survive in airshows and were used in the movies as Japanese zeros.

Last, and perhaps the most memorable was the C-47. The famous Douglas DC-3 went to war as C-47. The military version is distinguished by its large cargo door and barren cargo space. No interior furnishings.

Fold up benches for use by paratroopers. The civilian airliners were made into trucks.

I flew a goonie bird as third pilot, on flights of over eight hours. Being an unpressurized ship, we flew at low altitudes; 'Always in the weather. Without power assisted controls or auto pilot the physical demands could be exhausting. The particular ship I crewed was steam-heated and had extended range fuel tanks.

I just love hanger talk. While I did not have a lifelong career as a pilot, the romance lingers.

TO SEA OR NOT TO SEA

North Albany, the town I grew up in, was in the heavily developed area where the Mohawk River and the Hudson meet. Since Dutch colonial days its destiny was determined by the geology of the region,

The waterways provided transport and power. The soil made it a great farming region; surrounded by forested mountains that were rich in mineral wealth.

Even before colonial times, natives plied the north-south Hudson-Champlain valleys. The Mohawk Trail was trod from western New York, eastward over the Berkshires to the bay on the Atlantic.

The United States was born at the beginning of the industrial revolution. Animal power was replaced by mechanical means. Rivers provide cheap transport and abundant streams were available to power the many mills. First roads, and then railroads followed the 'water-level routes connecting the many mill towns. Canals, such as the Erie and Champlain canals were built.

The Erie Canal allowed barge traffic to travel from the Great Lakes to the Hudson, down to New York and the Atlantic; or up to the Champlain and Canada. The canal system proved very successful. Mule-drawn wooden barges were led by crews who walked along the 300-mile tow paths steeped and legend and song.

For the record, the waters of Lake Erie did not flow eastward to Albany. The water used to lift and lower the barges in the many locks was provided by streams and lakes flowing generally north into Lake Ontario serving several lakes ports.

With the introduction of steam tugs and steel barges, the old hand build dams and locks were enlarged. Please notice, each lock was built with a dam which pooled the water to the next level. Also, many dams had hydroelectric turbines generating electricity, powering industries and towns nearby.

Overall, it was a good system earning New York State the title 'Empire State.

Things changed dramatically after World War II. Most significant, is the St. Lawrence seaway, allowing ocean ships direct access to the great lakes.

Diversification and foreign competition took effect. The factories were antiquated and too costly to renovate.

The many smokestacks have no smoke. New York State was the most populous state – it is now in fourth place behind California, Texas and Florida.

A CHANGE OF DIET

Some creatures are carnivores eating meat as their diet. Vegetarians avoid meat and seem to live well. Most humans are omnivores – that is they survive on a varied diet.

I was born to a large lower-class family at the onset of the depression. My parents and their children survived on meager diet, consisting mostly of vegetables and bread but little meat.

In the age of iceboxes, before refrigeration there were no frozen foods. Availability of foods varied with the seasons of the year.

Spring was welcome with strawberries, asparagus, etc. Through summer, corn on the cob, various fresh vegetables lead to fall and apples. Farmers tedded hay, made silage, and prepared for winter. Swine herds were slaughtered. Peoples' diets changed with the seasons.

Old world methods of preserving food were still in practice after the depression and throughout World War II. Smoked meat, dried fish, pickling everything cabbage, onions, pig's feet, etc. The population was healthy, lean, and hard-working. Even during rationing programs, we enjoyed sports activities, and were relatively disease-free.

My family and neighbors seldom saw a doctor. The few doctors in town made house calls. Midwives and wet nurses still provided help, when needed.

Pharmacists made pills and elixirs, hand-printed labels and wrapped their products in neat papers. Hospitals were for surgeries. Sanitariums were for quarantines.

Frozen food from stores could be kept in refrigerators at home after the 1950s. Refrigerated ships bought food in off-season. Fruits and vegetables were available year-round. Labor saving machinery not only increased availability of consumer goods but allowed leisure for many people. Then people had time and resources for hobbies and entertainment.

Outdoor activities, such as canoeing, sailing, hiking, hunting became more mechanized. Roughing it in the woods, with scouts around campfires became less frequent.

Prepared foods, spectator sports, less dancing and mobility have certainly changed the lives of our nation.

Homemade anything means made from a kit. Sportsmen are now motorized; meals are from cans or frozen.

A comedian on television remarked, "If they outlawed can openers we'd all be starved to death!"

MY ONE LUNG SKIFF

I had driven past that overturned boat a dozen times. It was laying against the shed, with weeds growing over it. So, I pulled into the drive and inquired about it.

I expected the usual tale – that is, "My son is going to fix it." But, to my surprise I was told to take it away for 40 bucks. My brother, Bill Hill, helped me take my new prize to our yard.

The boat was about 14-feet long, built of good wood, fasteners, and marine hardware. It was an open fishing boat with a one-cylinder, air-cooled engine. We guessed it was used by a fisherman for many years. Not a pleasure boat of modern design.

The motor was installed well forward, with the shaft set for direct drive, forward only, to allow rowing.

A homemade two blade propeller thrusted against a pillow block. The stuffing box was also handcrafted.

My uncle, an old river rat, commented that both should be restored and kept in a museum! I'd like to have some fun with it.

A pair of oars fitted into oar locks allowed the boat to maneuver in close.

Steering was by a whip-staff on the port side, with rod linkage to a flat blade rudder.

An unusual feature was a copper tubing run the length of the hull. Curved from behind the prop, it delivered a stream of water into the exhaust pipe, which stuck out of the starboard side. That cooled the exhaust and deadened the sound.

A pie-shaped fuel tank was close fitted in the bow, holding about three gallons of gasoline. That could be a problem. We figured about ten hours running on a full tank.

You may forget to check the fuel. My brother once had a long row home, for such an oversight.

Speaking of problems, the engine had a serious one. The spark plug hole, in the cylinder head, was heavily damaged, beyond fitting a new plug. My brother commented, "That's why you got such a bargain."

However, I was being clever and thought of my oldest brother as an aircraft mechanic. We could install a 'Heli coil. This technique uses a special tap to make a thread and tool to install a coil providing a thread for a spark plug.

Except for the fact, the largest size Heli coil was much too small for the damaged hole. Gloom settled over us. The idea of refitting an antique, unsellable, engine was remote.

Just then my uncle Mike, pointed with his pipe towards his cellar door. In one of the several cigar boxes nestled between floor joists, were sparkplugs.

He picked a plug used in Ford model Ts. It had a green porcelain insulator and a tapered pipe thread. Using a plumber's pipe thread tape, the antique spark plug was installed. Under the guide of these masters, the engine came to life, after years of neglect.

A few final touchups, and we launched and on a fine summer day. A group of family and friends took rides in the simple, little motorboat. Over summer, my wife and I went boating, especially in the evenings. We left and docked up-river from the port where ocean ships came. My wife, of Norwegian descent, and I would read the sterns of freighters and tankers.

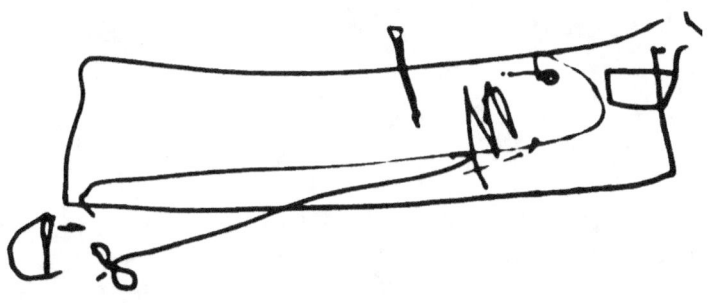

Once, when miles downriver, we were caught in our open boat, when a sudden rain while surprised us, as luck would happen, a friend in his Owens cabin cruiser drew near. We climbed aboard his sheltered boat and took our craft into tow.

Now, I had stopped the motor in our boat by shorting the sparkplug. But when we moored at our mooring place the motor kept running and the boat kept making way. Whoever heard of pulling a cord to start a motorboats engine?

My career took me away from home and the little putt-putt boat was used by family and friends. Over time, it was less used and neglected.

Younger people desired high powered boats for water skiing. The wooden hulled skiff needed loving care but fell into disrepair and uselessness. Eventually discarded, the little boat remains in our memories only.

PARADE DRESS

I remember getting ready and marching in military parades. Getting ready was a rather lengthy and detailed process. One didn't just get dressed and run out to meet the world. The first step began the day before by shining one's shoes.

In order to pass inspection, the process of spit-shining sure took time and procedure. First, the shoes must be cleaned with soap and water and dried. The application of Kiwi wax, then brushing and buffing was lengthy. To achieve the desired luster – especially the toes, required working the polish in with a rag. Failure to shine may result in a 'gig and could cause you to miss a liberty party.

The dress uniform was detailed the 'Uniform of the day' announcement. After having bathed, shaved, and checked one's hair, personal care items are used.

Donning clean underwear, black hose, held up by garters are used. Then comes starched white shirt studs to fit a paper collar and cufflinks for the French cuffs. High-waisted trousers held up by suspenders come next. A black tie in an Ascot knot, held by a tie clasp. Brushing off the jacket before donning it, almost completes the process.

The crown glory is the peaked hat. Always worn square on one's head, never like in the movies.

This piece of uniform attire is a very elaborate and expensive item of apparel.

The officer's dress hat is never worn in combat or on the bridge of a ship. It is for ceremonial and official requirements. For one thing, it is too expensive an item to risk being blown away.

After inspection, including fingernails, the units are formed up. Regimental staff with ceremonial swords and 'guide-ons' with their staffs and flags, begin the march onto the parade field.

One hopes the guest speaker isnt a long winded one.

*Parade Rest is a command used in the military to closely space troops information.

FIGUREHEADS AND MORE

Ships, in the great days of sail, had bowsprits projecting forward and up from their bows. They were held in place by cables port and starboard and down by a stay, usually with a dolphin striker. Fore-stays reached up from the bowsprit to the foremast carrying the jibs and foresails.

The thrust of the sails was transmitted to propel the ship by the foremast, held in place by the shrouds anchored in their eyes.

Perhaps the structural features were of little concern to people viewing the ships. The few sailing ships, now in museums, are photographed by visitors, especially the figurehead carried at the bow.

Woodcarvers delighted in making figures to be displayed on ships all over the world. Some of them famous. Such as Cutty Sark.

Cutty Sark was a ship made of strong oak, well-found and maintained. She served for many years and his on display in England, open to visitors.

The Scottish words Cutty Sark mean 'half-slip. The life-size figure of a young maid shows her naked from the waist up, her hips clothed by a slip, she holds in her outstretched fist a bunch of long hairs.

Curious viewers may ask who and why the ships name and figurehead came to be.

An explanation is an order. Scots love to tales, long cherishing myths in the lore of their homeland. This tale is of Rob Roy. Riding home on his horse after a night with the boys, he spied a bonfire in a thicket. Drawing near, he heard laughter and music. People were dancing and having fun.

He specifically viewed the most beautiful lass, with long flowing hair and her bare bosom. She turned and noticed Rob.

Their eyes met for only an instant. He realized he happened upon a group of witches and warlocks.

He immediately spurred his horse to leave. The young witch called Nannah, flew after him. The startled horse ran, frightened for himself and his rider. And off in the moonlit night they went.

Now legend has it that the witch cannot cross water in pursuit of her violator. Just as

---

Figureheads replaced by ornamental brass plates that were removed doing combat and later displayed in parks by groups of sponsors. Battleships were named after cities, and cruisers after state capitals.

they reached the bridge to town, she reached out in a desperate lunge and caught the horse's tail.

So, there you have it. Now when you are in a liquor store, you'll have a tale to tell – a tale about a tail.

KNOWING YOUR ROPES

For over half the history of the United States, sailing boats and sailing ships were the main means of communication and commerce. America became a great maritime nation. So, the population, specifically along our coastlines, knew a lot about sailing vessels.

Steam ships and motor vessels replaced sailing ships, but the skills and terminology remain for yachts and small crafts.

The common knowledge of sailboats and their rigging resides in a small group of people. First, let me make clear the use of the term rope. Rope is a word seldom used by sailors. It is the equivalent to carpenter's lumber. You should never say, "Hand me that lumber."

Rather, you should speak of the many lines of a craft by their use. Most sailboat have two general classes of lines, standing rigging and running lines.

Standing rigging refers to the shrouds and stays which hold the mast and bowsprit in place. Once properly organized, standing rigging requires little attention. Shrouds transit the power of the sails to the hull, driving the ship.

One notices that catlines, use as footsteps, going up the masts. Masts can be made of several sections stepped upward. The lower ends of shrouds are connected to eyes, a form of blocks, without sheaves pulleys. This method of staying a mast has been changed to turnbuckles.

Running rigging refers to the many lines used in operating the sails, requiring constant attention and adjustment. Each line has an individual name. When properly used, there will be no confusion.

The simplest of sailboats, such as a sailing dinghy, has one sail. Fixed in place by a top halyard and an outhaul, it is regulated by a sheet. Single-handed, the sailor keeps one hand on the teller and the other on the sheet. Letting go of these controls the boat will point up into the wind and stop. Ships with many sails require much study and familiarization.

Square rigged vessels are today mostly for 'barefoot' cruises.

In the days of the clipper ships, many boys were required to go aloft to manage the sails. This was especially dangerous in heavy weather.

Sloop-rigged vessels, with sails rigged fore and aft, are manipulated from the deck, seldom requiring anyone to go aloft.

Sailing ships now in paintings and movies, are usually depicted in ideal conditions and viewed by those who know little about them.

The few ships in museums are disappearing leaving sailing ships to model builders. They take pride and spend much time and attention to details. God bless them.

MY U-2 STORY

Doing the buildup of the Eastern Test Range, I worked as a design craftsman at Melbourne regional airport. My efforts concerned making drawings and working closely with shop personnel producing the prototypes of camera locations being built down range.

The cameras were provided by the Air Force were Perkin-Elmer units removed from RB-36 aircraft. Our group was tasked with providing rapid build and install of a complete system at perhaps a dozen locations. With a small engineering group and a precision machine shop we were a 'turnkey' team.

Our location at Melbourne airport was within driving distance of Cape Canaveral and Patrick Air Force Base.

Built doing WWII at Indian River Naval Air Station, it was home to planes that shepherd convoys up the Gulfstream. It is said that never was a ship lost to a U-boat, while escorted by a blimp.

The circular landing pads where mooring masts that held the ships could still be seen. The several hangers and buildings were leased to a couple of firms.

Fellow workers soon learned of my interest in the airplanes and that I was there with an outfit that built twin-engine aircrafts nearby.

So, on a break, I checked it out. Expecting to see many workers at an immense array of machines and fixtures.

I introduced myself to a worker. He continued to drill holes at a drill press while telling me about their aircraft business.

They modified single engine Ryan 'Navions, removing that nose motor and prop, then attaching a motor prop to each wing. He informed me the rest of the workers were out to lunch – all three of them. He commented, "The four of us have the Wright brothers outnumbered two to one!"

It became a habit for some of our workers to walk about the airport during lunch to stretch our legs and see the sights. One day sirens sounded, and vehicles came out to the longest runway. Ordered to stay well away. We were told an aircraft was approaching with no power. Off in the distance the silhouette of a ship was making a very shallow final approach. What appeared to be a soar plane landed on bicycle landing gear and came to a halt leaning on one wingtip.

Vehicles from Patrick Air Force Base immediately sent crews to take charge. The canopy was opened and down the ladder came to pilot Wearing a G-suit, spurs and helmet. He was whisked away in a staff car. Outrigger wheels were placed beneath the wings, leveling the dull black Aircraft, which had a few markings identifying it as USAF.

News reports explained that happening, stating that it was an air force U2 jet plane. Designed to fly at high altitude doing upper Atmosphere research. On this mission, the U-2 ran out of fuel halfway from Bermuda. Normally a landing at Patrick would be accomplished. But a storm cell moved onto the runway. So, the craft continued its approach to Melbourne runway.

The airplane was refueled and ready to fly across the Indian river to the Air Force Base. Another pilot, in flight gear was strapped and ready for takeoff. A power generator spun up the jet engine. It ignited and the preflight check list was completed. A surprisingly short run and a steep climb out left the tip wheels bouncing down the runway as the U-2 flew away.

That incident attracted little attention and quickly died, being of local interest. However, U-2 aircraft became a worldwide story when Francis Powers lost his ship over Russia.

Being involved in things aviation, I sustained an interest in the ship itself. Having worked for Lockheed, I was aware of the 'Skunk Works, and Kelly Johnsons' products. Also, while at Kennedy Space Center, Cecile Powers, Francis' brother, lived in Melbourne and, over time, the super-secret became known.

The U-2 aircraft was a jet powered soar plane. Soar planes and gliders do not have motors. So, rarely, aircraft are introduced that can be launched under their own power. They should be called motor-gliders or a motor-sailplanes.

Usually of light weight, they tend to be expensive, using high-strength special materials. Soar planes are characterized by long thin wings.

They are known for a low rate of sink. Say, from an altitude of 5,000 feet above the ground, a light plane could go ten miles. A glider maybe 20, and a soar plane perhaps 40.

An accomplished soar plane pilot seeks out uplifting air currents and gains altitude. A 'Silver C Award is earned by soaring a cross-country, triangular, course. One leg of 100 miles, one leg upwind.

The U-2 spy plane operated in the upper atmosphere above other aircraft and most anti-aircraft offenses with a highly modified jet engine. The compressor sections first stage of the motor was much larger. This allowed the unit to gather a lot of thin air into the jet.

Fuel is lost to evaporation, More so as altitude increases. Special fuels were deployed to reduce losses.

I guess you might call me an anti-gravity nut. Since childhood, I've been ready to go above the earth whenever and wherever. Model aircraft building kits of balsa wood and rice paper marked my beginning interest.

Piloting 1930s aircraft and up to multi-engine planes, the romance of the sky lives on in me.

A U-1 aircraft is a Cessna (not 195). A single engined 180 hp Jacobs radial all metal four place, high wing, tail-dragger, utility.

A "U-1" aircraft is a CESSNA 195; a single engined (180 hp Jacobs Radial) all metal 4 place, highwing, tail-dragger, utility.

BOEING P-26

A BIG SECRET - HOW WE WON WORLD WAR II

As a schoolboy during the 40s, I really couldn't gauge the magnitude of the historic events taking place all over the world. Mostly, I speak of events that I recall in later years, from what I read or heard from others. My personal involvement, which stays in my recollections, centers around my family and my father's business.

Anthony Hill, my dad, was a self-employed welder. He owned a small fabrication shop, making mostly war goods. You see, we lived close to the Watervliet Arsenal, an army facility where guns and other goods were made. An order had come out for all shops in the area to register with the defense mobilization board. All the best businesses were visited, by inspectors and the capabilities assessed. Work orders were issued, and plans put into action so many shops contributed to the war effort.

With our garden variety of mostly used machinery, my dad's small place employed a few elderlies, retired railroad men, myself, and my brothers. Labor was in short supply as all able-bodied men were in the service.

Authorities looked the other way when underage youths were put to work on farms and factories.

The processes of making many different small parts began. Not even knowing they use of certain items – better not even ask, we cut, bent, drilled, forged, and welded, grinded and painted all sorts of small parts and assemblies.

We could only guess how the use of thousands of guards for headlamps, grab irons, machine gun mounts, brackets for bunk beds and antenna mounts. We worked into the night and after school and weekends, alongside the adults.

The entire population was behind the war effort in any way possible.

The many factories, large and small industries, the port areas, as well as rail and highway activities concentrated at the juncture of the Hudson and Mohawk Rivers were considered prime targets for enemy bombers. Especially those working at night.

This gave rise to the need for blackouts. I later heard civil defense authorities used the blackouts for psychological as well as military needs. This national emergency brought a peculiar arrangement right into our family.

My father's business was licensed for the storage, use and sale of cold compress gas cylinders, such as oxygen, acetylene etc. He also was licensed as a detonator and trafficked in dynamite.

We stored dynamite and caps and fuses in a concrete vault in a lot behind our shop. Normally we sold and used the dynamite for stump removal, opening graves and such.

I can remember, we had to turn over the boxes in storage, every so often. My dad no longer handled the sticks, but rather my brothers and I were his powder monkeys.

We went out to road building sites, when needed to blast rock. These detonations were not like in the movies. There were no giant flashes, loud bangs, or huge showers of rock. Rather, the contractor drilled holes in the rock and filled them with water.

Our job was to dry the holes, placed the charges and connect to the magneto box. The blast area was covered with a blanket of woven steel cable as a debris screen. A thud and vibration shattered the rock, so it could be handled by heavy equipment.

During the war years, road building was curtailed. Then an unusual use for a detonator appeared.

My father was contacted by Judge Owen Conlin, who was appointed head of the local civilian defense activities. He had a need for my dad's exceptional skills. We would operate the blackout signals. Special equipment and their use had to be built and readied for use.

To this day I could make a signal projectile called a salute. It consisted of a three-foot dowel with a dozen or so cardboard washers stacked at mid length. Two sticks of 40% dynamite were taped to the upper portion of the stick and two half-sticks attached to the lower section.

A timed fuse connected the top and bottom charges and fuse to be lighted it was run a few yards away. The made-up salute was loaded into one of several tubes in a rack fitted into a concrete vault.

A coded telephone call from the judge alerted my father and we prepared to act. Two hardened locks were removed, and the counter balanced steel cover was raised. At the prearranged time. Ignition of the lower propellant charge set the missile aloft with a loud thud.

Hurled a few thousand feet into the air, the upper (signal) charge is made of bright flash and loud boom. The sound reverberated for miles up and down the Hudson Valley.

From our vantage point we saw all the lights go out in the neighborhoods and towns. People darkened their homes with drapes and air raid wardens enforced the rules. On one occasion, the judge flew in a light plane and took photos of the area.

The pictures appeared in the newspaper and violators were readily noticed.

The war ended when I was 13. The equipment was no longer needed. The helmet, arm bands, gas mask, etc. were stored in a downtown jail cell or discarded.

With little fanfare, we took the explosive charges across the railroad yards to a large sand pit. The sticks were cut with a Barlow knife and laid on spread out newspapers. When the paper was lit there was no boom – just a furious bright flame and smoke.

The war was over. Certificates were presented to the team.

That was 70 years ago. No need for secrecy now.

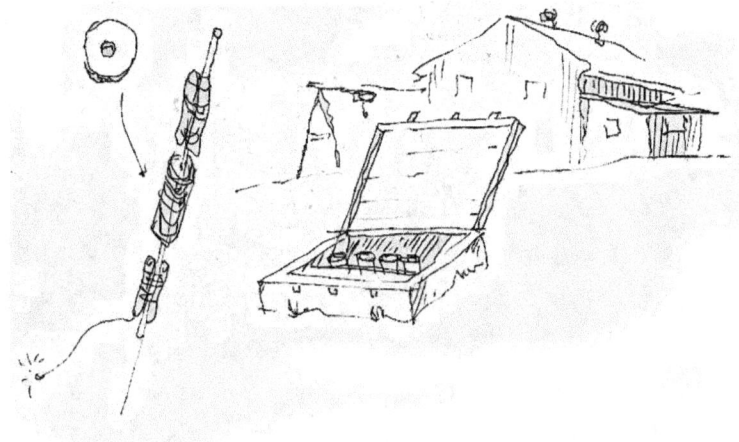

HOLE IN BLOCKHOUSE WALL

I was one of about 100 volunteers at the USAF missile and space Museum, located at Cape Kennedy Air Force Station. Normally, my duties include acting as docents at launch complex 26 (LC-26) for two or three days a month.

One occasion was a special open house for space workers and their families. The guests were allowed access to several launch complexes following at tour route.

My partner Ron and I were detailed to launch complex 14 the Atlas Mercury pad where John Glenn made history piloting Americas first orbital flight.

There is little to see or do now. Perhaps the stainless-steel monument with a buried time capsule, makes a good picture. The launch pad itself has been demolished and so overgrown that it is it practically inaccessible. The Blockhouse is a rather limited attraction.

There is little to see or do now. Perhaps the stainless-steel monument with a buried time capsule makes a good picture. The launchpad itself has been demolished and so overgrown that it is it practically inaccessible. The Blockhouse is a somewhat limited attraction.

The Blockhouse is a giant concrete igloo. The entrance, from the west, features a massive thick steel door.

Inside the Blockhouse, all equipment was removed, and the space was carpeted. The room is used as a meeting space with a suspended ceiling and fluorescent lighting and furnished with tables and chairs. The few pictures on the windowless wall are of little interest. A Spartan offering indeed.

I bought a long book of cryptograms to while the hours.

However, this must have been a slow day throughout Central Florida. Large numbers of autos and vans, filled to capacity, started to arrive. We lost our headcount so, we estimated between 800 and 1,000 visitors. Once inside, they went directly to the three periscopes, used to view the pad area. There are four adjustments on each periscope, I know three scopes were manipulated by each visitor, making them useless and of no interest.

The main attraction, a surprise to Ron and myself was the air shaft. That is, a hole in the Blockhouse wall, opposite the entrance. Approximately three feet in diameter, the shaft was fitted with light bulb. We had intended to show the eight-foot thickness of the concrete at that area of the block house.

Incidentally, the shaft was fitted with a set of ladder rungs, leading up and out. The designers intended this as an escape route in case the massive blast doors couldn't be used in an emergency. However, in recent years the outside and had been blocked, to keep out varmints.

A boys curiosity got the best of him. He scampered into the hole and in short order was followed by others. The girls, not wanting to be outdone, entered. Their followed fathers, mothers, grandparents. Even a toddler, unable to walk, tried to go to the length of the tunnel.

People were standing five or six in a row, awaiting their turns. This turned into a photo op. Everyone seemed to have a digital camera or a cell phone that took pictures.

Our visitors, some in groups, had us take their photos. Ron and I were amazed. He commented that tunnel may have been dusty and cobwebbed but it was sure clean now!

At the close of the open house, we drove back to the museum office at LC 26 to check out and go home. In trying to understand that phenomenal activity I reflected that I had taken several courses in psychology at the University, but my study of human behavior is not complete.

At best, I relate this activity to the practice of going to Ireland and having a stranger hold your ankles as you bend over backwards to kiss the Blarney Stone.

As an afterthought, our budget minded Director said, "Perhaps we should have put a donation box there and collected a few coins."

We ended the open house day, with the comment that neither Ron nor I were inclined to go into the hole in the Blockhouse wall.

WATERVLIET ARSENAL

I worked as a draftsman at the Watervliet Arsenal for a short time. It was Americas oldest armory and largest gun manufactory. This was to be my career.

But, fresh out of high school, appointed as an apprentice tool and gauge designer, my joy turned to woe. Here I was in this historic place; temporarily. Might as well make the most of it and pray things might change. Weeks turn into months, and I did have a valued experience. What I learned there carried me throughout my working life.

The arsenal, located on the upper Hudson River, is a huge place. The US Army, over the years, has manufactured some of the largest guns in the world. With barrels up to 140 feet long, and a bore of 16 inches. They were used on battleships and at coastal defense locations.

Fifteen thousand workers toiled 24/7 during World War II, turning out a variety of munitions pistols, rifles, mortars, machine guns – you name it.

However, the large guns were no longer needed, and smaller arms were being selected by our allies in NATO. The mission was redefined.

In the event of an emergency, the army was to be prepared to provide many thousands of weapons for our forces and our allies. This meant, tooling up and beginning production overnight.

Even with the use of civilian providers of parts, time was of the essence. Design, produce prototypes to be tested, and provide for mass production followed in steps. Upon a design being selected, a small lot of prototypes would be handmade by the most skilled machinist.

The Watervliet Arsenal was originally in the sparsely built-up area near the Erie canal. They could test the weapons of a century ago on the property. The armory expanded over time and the entire area was built into a dense, heavy-industrial region.

The Watervliet Arsenal then became the landlord for the Aberdeen proving grounds in neighboring Maryland.

Guns and equipment to be tested were sent to Aberdeen and proof-tested to destruction, evaluated and approved for production.

Production required all sorts of items. Fixtures to hold parts being assembled. Jigs and drill plates for match drilling, punch and die sets for use and punch presses. The list was very long and impressive. The tools must then be provided.

A short run of finished product will be used in the field. Several sets of tooling will be ready and placed in storage, ready for rapid deployment.

I learned a lot of things not taught in classrooms. Being with skilled and specialized engineering staff members was a great experience. I was sent to places in the sprawling shops and offices; amazed as I wandered through the facility.

The huge 'Tube shop, referring to the place where large rifle barrels were made was awesome. Lathes, over 140-feet long, able to turn eight-foot diameters were there. Broaching machines pulling a 40-foot brooch through a 140-foot-long barrel had to be 400-feet long. These rifle barrels were machined and annealed when in a soft condition. To be heat-treated and toughened, they were moved to electric-heated 'Shrink Pits. These ovens were at the end of a high-bay shop.

This place is huge. Railroad cars move in here, amongst large machines, with several bridge cranes above. A 140-foot gun barrel would be heated and shrunk in a manner done by armorers for centuries. These tubes can't be laid horizontally on perches in an oven.

They are listed vertically, above machines and rail cars and lowered into pits almost 200- feet deep. Add it up – a 140-foot tube, 20-feet above the floor, a crane of 20-feet and the roof trusses of maybe another 20-feet. Add it up the shop was about 200-feet high and a good 1,000-feet long.

An aerial photo will show many buildings of various shapes, and sizes, connected by roads with parking areas, and railroad and trucking terminals.

I grew up amongst industrial areas and was familiar with general layout of plants. So, one day my Lead (boss) send me to meet with a man who was in the Gray Iron building, room 202. Gray iron is the term used for cast-iron.

Naturally, the foundry would be at the rear of the site, near the rail sidings. Upon asking for room 202, I was told that there was no second floor. Bewildered, I returned to my office after a long hike. "Come with me," my boss said.

We left our building and crossed the way to a big building nearby. He banged on the wall and said, "This is the Cast Iron Building."

Sure enough, this very large structure was made of ornate cast-iron panels. They were cast right on the site and erected, around the Civil War time. As I learned, it was not uncommon. As a matter of fact, the dome on the nation's capital, is the largest cast-iron structure in the world.

Another remembered set of tasks, involved the Walker 'Bulldog army tank. I was never in a tank before and went to the shop while we were making several changes to be evaluated and, maybe, incorporated. This tank was painted white enamel inside and kept clean.

I was involved, in minor ways, with two assignments. One was installation of radio in the command tanks. The other was tooling for an improved main gun. Let me explain

Most tanks have short range radius, being able to talk to each other, not to be heard by the enemy. The lead tank has the capability of talking to its combat command center. Many rounds of ammo are stored in the turret of the tank, ready for use. Command tanks have a bigger, better radius, but fewer rounds.

In battle conditions, there is much banging around, hence the need for helmets. nerf bars are installed to protect the radio. The cage of steel bars protecting the set must also allow the knobs and switches to be operated by a gloved hand. The drawings I had approved were used to make a bunch of installations for evaluation.

When viewing the tank from the outside, the large cannon has a cylinder beneath the barrel. As a boy with model toy tanks, I thought this was a shock absorber.

It serves a different purpose. If you had seen real tanks firing the main gun, the blast of the shells fire and smoke, may be followed by a donut shaped smoke ring.

This is caused as follows: when fired, the projectile travels from breach to muzzle. Holes, called gas ports, located halfway to the muzzle allow gases to go the cylinder below. When the gunner lowers the breach block for reloading, a valve allows gas from the storage cylinder to blow the rifle barrel out, preventing gases from coming into the crew's space for their safety.

Another feature, the smaller rifle gun alongside the big cannon is not a machine gun, to protect the tank. It is a special single shot rifle. It is fixed to the canon and moves with it. Unique ammo rounds are kept for its use. These bullets have the same trajectory as the main gun. A typical use would be in combat.

Tanks support troops would be positioned and made ready for target like a bridge. The gun aimer points at the point on the side of the bridge. The special round is fired. The gunner, thereby, gains information that allows the next round to be right on.

Another novelty concerns the Arsenal working 24/7 during World War II, even during blackouts. I noticed the shops were lit with sodium vapor lights. All the windows were tinted blue. This combination was found years later while I was at Emory Riddle.

Let me explain. In pilot training, students went 'under the hood,' to learn blind flying. Actual flying, with a canvas cover above the Link, vacuum-powered simulators. In modern aircraft uses for instrument flying lessons, the glass areas are yellow, and the student wears blue tinted goggles. The instructor, a safety officer sees very everything in yellow. Students see the instruments and controls in blue. But the windows appeared totally black. I thought of those lights and windows at the shops in Watervliet Arsenal.

The arsenal is maintained by a skeleton force, ready to jump when called to do so. My experience served me well, as my non-career took me to other tasks in other places. I'd do it again.

HARDSHIP DUTY IN NEW YORK

Sampson Air Force Base was named after an admiral. It was built in 1943 as a Boot Camp. Newly enlisted sailors were given military basics and expected to spend about eight weeks there.

The buildings were expected to last three years. In 1948, the Air Force was separated from the army. I was stationed at Sampson in 1955 and 1956. Nobody, I mean nobody, enjoyed being stationed there.

The site in the Finger Lakes region was subjected to severe weather, being in direct line for polar storms coming across the Great Lakes.

The barracks were light framed, uninsulated, and two storied with open bays.

Set upon brick piers, two feet above ground, they were heated by hand-fired coal furnaces. A sheet metal duct, going down the center of the lower bay was woefully inadequate.

Heating specialists (stokers) were assigned several buildings in which to maintain the fires. They were issued arctic clothing to make their rounds. A common complaint was that the tenders overfilled the fire boxes.

I recall, placing a small rug on the floor near the bunk, to step on in the morning. The strong cold winds coming up from the lake would go into the barracks, through the floor cracks causing my foot rug to move to the wall, far away.

These buildings were tinder, quickly consumed when caught afire. A standard joke was in the event of fire, walk, don't run, through the nearest wall!

The only steam heated solid brick building on the base was a stockade and jail. Finally, orders came down to close Sampson.

The air surgeon general (top doc) stated, upper respiratory disorders have caused delays and disruption of the entire training command. Henceforth, boots will be sent to Lackland, in Texas or Parks, in California.

A lot of people were jubilant. But there were some down sides to the closing. It wouldn't happen overnight. The training group graduates 80-man flights every two weeks. This process would empty the ranks of trainees. When the instructors had no more students they would be shipped out. That would leave the airbase group, with a base without a mission.

Air bases are organized along two-tier systems. The airbase permanent party mow the grass, fixes the plumbing, maintains security, and generally does all the housekeeping of a base. As a host for a mission-group, bases from time-to-time rotate missions. Bombers, fighters, transports, come and go as the needs of the Air Force reveal themselves. Bases are closed or reopened.

While Sanderson was in full swing, the abundance of 'boot-power was very useful to permanent party units. For instance, send a dozen 'basics on a garbage detail, or to mess hall duty, or to clean the windows.

Suddenly, duty schedules were made for all personnel, regardless of rank. Can you imagine a six-stripper (master sergeant) riding a Dempsey dumpster? You could learn new words.

The hospital received no new admissions and was evacuated. A big problem was what to do with the equipment and inventories?

Word spread fast, and the scavengers came. The orderly disposal of government property was carried on, for a while.

The single runway at a basic training base normally has little traffic. The tower was not manned around the clock for years. Runway lights were used upon request. Things change, and the vultures descended.

Once the training function ceased, the base had no mission group moving in. The general took his flag and left.

The wing now became a group having few squadrons. The closing continued. Buildings were abandoned; personnel sought positions elsewhere and many 'top graders retired. The base took on the guise a ghost town.

I wanted to go to tech school, become a jet mechanic, electronics tech – or anything. But the needs of the Air Force were lessening. I would have liked to go overseas on occupational duty, but no one wanted an airman with my qualifications. At best, I could stay n the training command.

Well, the military does everything by the book. When being transferred or discharged, there was a pile of paperwork going to interested units.

First, you must have orders cut – permission. This begins your 'out-processing. It usually takes four days. Then time for travel ten days to California. Then in-processing at the new base required even more time. About 20 days to go to new base.

I had never taken a leave. I accumulated 60 days. If upon arriving at my new base, I put in for discharge. That would start the business of processing all over.

So, after many tries and rebukes, my squadron commander said, "Why don't you just go home now?" I thought it over for 15 seconds.

I went home with over 60 days-pay, an honorable discharge and I satisfied the draft in 22 months – not 24.

P. S. Permanent duty personnel worked 12 to 14 hours a day. The law says, 'persons are sentenced to hard labor are to be worked a maximum of six hours!!

P.P.S. My squadron commander was valedictorian of his law school class. He wanted out, but he had to counsel me about the advantages of staying in the Air Force. Being young, and not too smart, I said, "If they start World War III, I'll be back. But don't start a war to get me back in the service!"

FIRE HYDRANTS AND MORE

I was hired as a design draftsman to work with a group engaged in modernizing the works of the largest manufacturer of fire hydrants. One might imagine a production of several dozen a day might satisfy demands. Surprise! – they make truckloads every day in several plants. Fire hydrants not simply valves.

Assembly drawings reveal a long list of parts. Hydrants are a rather complicated and expensive product. The materials used are stainless steel, bronze, cast-iron and high strength hardware and special gasketing. Ask an insurance company, what replacing a hydrant might cost.

By the way, when struck, hydrant does not shoot volumes of water skyward. There is a feature called a 'Raffic fitting located way down, that prevents loss of pressure in the main.

It allows other hydrants in the system to not lose pressure. The only hydrant that squirts up to the sky is in the back lot of a movie studio. Another device incorporated is located at the bottom is a relief valve. When the hydrant is closed, the water in the stem is drained, so the hydrant is not subject to freezing.

The variety of sizes and features provides many options.

I was assigned the redesign of core boxes in the foundry. As usual, on large projects, teams of engineers and their helpers are assigned sections and milestones to be met. Project management closely coordinate time and money.

Foundry practices include pouring hot metal into a mold. The shape of the object being molded is dictated by a pattern. If a hollow item much like a valve body, engine head or exhaust manifold is to be cast, a core is needed. That was my concern.

The space in a hollow object is made and a core box. Specially mixed sand is packed in and roasted. When properly made the core is much like terra-cotta used in flowerpots or barrel roof tiles.

The outside shape of an object is dictated by a pattern used by packing molding sand into a mold. This flask is usually two halves, a cope and a drag. Arrangements must be made to hold the core in place. Wire and perches prevent the core from floating out of place. After the pour is cooled, the casting is removed. The sand is reused, and the object has its core removed. Those so-called freeze-out ports are used for emptying the core from the cast item.

For centuries, cores were molded in specially made boxes filled with a cementitious compound and allowed to dry. Removed from their molds, they were then baked in a core oven (kiln). This was slow and labor-intensive.

My team designed and supervised a modern, electronically controlled, high-speed making of cores.

I should mention that we were expected to be on this location for, perhaps, two years. Shirley, my fiancé, and I began plans for marriage. With the help of our families, plans were made, and bans announced. All the fixings of a classic wedding preceded.

Shirley and I planned to honeymoon in Florida and return to upstate New York. I had worked in Florida first and she wanted to see for herself and meet my friends.

Disaster struck as the wedding date approached. I was being terminated – out of work!

The ceremony in our old church was beautiful. The wedding supper and dance was marvelous. We headed south and I remarked, "Theres no rush to get back."

So, we drove our Mercury to Orlando. With rounds of introductions and socializing, we had great fun together.

A friend of mine, and his new wife, was kayaking in Indialantic. The next day was a Sunday. So, we agreed to put on our bathing suits and meet on the beach.

After hugs around, I was asked, "What are you doing now?"

I replied, "Im out of work and looking."

My buddy immediately turned and yelled to a man lying on a towel nearby, "Hey, Jean. Heres a designer available right now."

We talked and he hired me on the spot. The contract was development of camera sites for the eastern test range being developed. From that day on the beach, I was never out of work on one space program or another, all the way to retirement.

Who writes this stuff, anyway!

P. S. I did not include the attraction that hydrants have for dogs!!

TOBACCO THEN AND NOW

In 1954, I was sent to Stallings Air Force Base for pilot training in Kingston, North Carolina. At that time Kingston was the world's biggest bright-leaf tobacco market. Several block long, red brick warehouses line the main street.

Once, out of curiosity, I went with buddies to witness a tobacco auction. The tobacco to be sold were arranged in neat piles of 'hands' set-in trays on pallets. All the tobacco was large, blond Burley, properly cured and inspected for blemishes.

We were told, tobacco was a major source of revenue for both federal and state governments, and as such, was supported. Prior to the sale, a government agent placed a paper on each tray. It had a price the government would pay for that consignment. As the bidding began, the auctioneer, in a classic blather of talk, walked briskly along, followed by buyers from many tobacco processors. As each sale was completed, the price was revealed. Most of the tobacco was bought by the government, to be stored and sold later.

Tobacco as a crop is unusual, in several ways. A single teaspoon of seed will make hundreds of acres of plants. But as the soil is depleted, tobacco fields needed to be rotated. It is very labor-intensive.

A system of tobacco allotments was used. The idea was to have each small farmer, have an acre, or less, to provide a minimum income to keep the family homestead.

Each plant had its blossom removed by hand, allowing sunlight to enter down into the leaves. Harvested by hand, the bunches of leaves were hung on long sticks in the curing barns.

A note on curing tobacco. A dried out and brittle leaf is worthless. The large, blonde unmarked Burley is bought for its looks. It has no smell or taste when smoked. It is mixed with other tobaccos and spices to make distinctive blends, for smokers.

Fortunes were made in the tobacco world. Advertising and promotion made smoking a popular habit.

Cigarettes were even sent to veterans' hospitals and given as treasured gifts. I've personally bought cigarettes at sea stores, for $.80 a carton. That's eight cents a pack.

Well, things have changed in recent years. For one the tobacco allotment for small growers can be 'assigned. That means people get money and corporations can use machinery to grow large fields of tobacco, with little hand labor.

Plants are grown to maturity and harvested by machines – the entire plant. Gone are the curing Barnes that dotted the landscape. Air cured tobacco is unheard of.

Steel mesh wagons of entire plants are rolled into sealed chambers. Electronic controls for heat, air and humidity prepare the tobacco for the next process. Using paper making technology, tobacco plants are ground entirely, made in to pulp and are now homo-generated tobacco leaf. (HTL) It looks and feels exactly like paper from a brown paper shopping bag.

Old tobacco was 'Navy' cut. That is, cut and crimped into uniform pieces, which kept lit and burned uniformly. Then made into cigarettes. The

HTL is shredded and made into cigarettes but may have all sorts of debris moved downstream to the smokers' lungs.

There were great savings, especially in labor in the field and in the factories. But threat of health problems and lawsuits were avoided by the addition of a filter.

Making a filter and attaching it to a cigarette was another mechanical device added, but still, good profits can be made.

And another consideration kept under wraps, is the effects of the gaseous contaminants which are inhaled.

In recent years, programs to reduce and eliminate smoking have been introduced. Non-smoking areas have increased, laws limiting sales and non-smoking products are, at the very least, used to discourage smoking.

Once, some bales of Turkish Latakiy tobacco were ruined by fuel oil spilled on them. The insurance paid for the damage bills.

I asked, "Will they destroy those bales?"

"Oh no," he replied. "We will sell them to Black Flag to make insect killer."

Epilogue

Way back when, a French scientist named Jean Nicot was an early player in the tobacco story.

Nicot observed the native Americans smoking cheroots, made of tobacco. In time, he he proclaimed the tobacco to be toxic and should be banned. His name lives in history as nicotine.

Centuries later, the comedian Bob Newhart made fun of Sir Walter Raleigh shipping a boatload of leaves to England.

CHANGING FROM CHAIN
(Product Designer)

I was working for a manufacturer of timing chains for automobiles when the decision was made to use rubber timing belts on auto engines, thus eliminating our main product. The second important product was carding chain used on textile machines.

Carding chain was in to be made in North Carolina thus gutting the plant in upstate New York where I was a designer in the engineering department.

Several meetings were held to determine our future. It was decided to come up with a variety of products to make use of our facilities and staff.

During the next months, we designed a family of ocean-fishing reels, an improved animal trap and an electric golf cart. Our small group was working for survival.

But the most fascinating project was to design and build the world's highest capacity ski lift, in Vermont. From our plant in northern New York, parts were to be made, tested, sub-assembled, and shipped to Vermont for installation.

This required occasional trips to the construction site, for measurements and consultation with the builders. We enjoyed the day away from our office to the open-air mountain side.

The final approval of the ski lift was with the Vermont Railroad Commission. They were responsible for all people-movers, such as elevators, escalators, amusement rides, etc.

We understood that the double-chair ski lift was a transportation system and insured to be safe. That is, paying customers, with tow tickets, ride to the top of the mountain and return. If people decide to leave the lift and ski down the slope and trails that's their responsibility.

The ski developers own the lodge and the lift. The slopes and trails are leased from various landowners. In fact, you can park your car, walk up a slope and ski down repeatedly. You can go inside buy a warm drink and not be challenged.

Back at the factory, progress was proceeding apace, to be finished and ready for the coming skiing season. Our facilities included a foundry used over the years for casting gear boxes and custom machine frames.

The most ambitious operation was casting and machining two giant sprockets for the lift. The 13-foot diameter sprockets required a floor mold.

That is, a hole in the dirt floor was dug several feet deep and 20-feet across. Specially molded foundry dirt was used with a segment pattern to prepare for the pour. The blast furnace was charged. When ready the furnace was stabbed, and hot molten iron ran across the foundry floor. Channeled into the open-faced mold.

Foundry workers tended this dangerous work to ensure it was correctly accomplished. The casting was dug out days later and sandblasted and cleaned – ready to be machined.

These sprockets, too big for normal machining facilities, had a specially built jig to accommodate the four-inch chain links and pins.

A bicycle chain would have links to be held between your fingers. This lift system had chains a mile long, each link weighing several pounds. Delivering these outside items required special permits and experienced freighters.

The completed tower, hangers with double-seats spaced the length of the system went into operation.

The most attractive lodge was built in A-frame style. With a huge copper smoke hood above the stone fireplace at its center, it was impressive.

Large glass panels at the end gave a view of the lift and slopes. A swimming pool had access inside and out. Everything was first-class, hotel, restaurant, and staff.

A sidelight to the project came to our attention. Several days before opening to the public, Playboy magazine had a photo shoot. Complete with centerfold models in and out of the pool. Unencumbered with attire, they spent the day there.

We asked about the activity, but the project manager was sworn to secrecy. He suggested we buy a copy of the magazine.

MY LAST SAILING RACE

Sometimes you can back yourself right into a situation. It may not be a good idea to brag about your past achievements and add an amount of hyperbole. Let me tell you about one of my such experience.

You see, I grew up around boats. As a young man I preferred sailboats over motorboats and spent some years at sea. My stories grew with age and people may have considered me an old salt. Well, that was long ago.

I recently got a phone call from a friend who had recommended me as a crewmember on an upcoming sailing regatta. He said, could expect a call from the owner of a nice Hunter cruising sailboat. The owner was short on racing experience, but he had lined up two other young fellows who claimed to be of recent experience. I figured it would be nice to spend a day relaxing on a fine yacht upon the briny.

On race day, all met at dockside, and exchanged pleasantries. The owner was dressed all white. White shoes and socks, trousers and shirt and a fine cap. He would make a phone picture.

I wore khaki shorts, canvas shoes no socks, Tee-shirt, brimmed hat, and sunblock. I expected to get wet and sweaty. One of the younger fellas didnt show up but the one that did appear was all gab about his knowledge and abilities.

We waited a while and then decided to sail shorthanded. We would make sail after getting underway. The owner asked me to take the helm, while he and the boy made sail.

It appeared to me, that the rigging of the sails and sheets lines was done mostly by the owner.

This was a fine boat. As we left the river, we went through the canal, bridges, and lock, to the harbor. The Yamaha diesel was running to help move the boat and it steered well. The owner appreciated that I was handy with the wheel, the engine, and signals.

We shut off the diesel, when going through the harbor and lined up with other cruising-class boats near the starting line, just about on time.

This boat and four others were classed as cruisers, being used for living aboard and not rigged for racing. As the slowest group, we were sent off first. The pure racing classes would be sent off later.

So, with a little fanfare, the canon was boomed. It sounded more like a pop. We headed south paralleling the beach. The morning breeze was offshore and with Genoa sail pulled tight, we were blasting along at a good four or five knots – a fast walk.

It took us several hours to reach Patrick Air Force Base. The morning breeze became the noon calm. The course now was due east, away from the shore. With a little wind, I rigged a whisker pole so we could go 'wing on wing downwind. I was told the cruises weren't allowed poles. This race was under NYCYC rules.

Did I mention, the captain had been at the wheel all the time? By this time, he was disenchanted with a young fellow who appeared appear to be a guest and of little help.

So, the skipper mentioned to me to take over, while he went below. The duty of steering a sailboat is quite different than a motor vessel. The constant strain holding a rudder and the pull of the mainsheet (line) required constant muscle use.

Keeping an eye on the sails, maintaining course, and watching the other boats is demanding. Pressing the boat to win in a race is not like a laid-back leisure sail.

The skipper was below for quite some time. I suspect that he used to head and perhaps had a rest and some coffee (fortified). Finally, I took the initiative to tell him we were at the change in water, the Gulfstream.

We now were headed north back to the committee boat, miles ahead. With a few knots assist from the Gulf Stream, the shore wind picked up and was steady. The boat was now healed well over. We were driving hard to the finish.

The hull smacked into the seas and sent spray over the deck. The helm was a tough pull as the lee rail went into the water a few times.

The owner appeared, with his binoculars, searching the other boats. All the catamarans and racing monohulls had long since finished. Of the five boats in our company, two were ahead and two behind us. The two followers sheared off, went directly home, and did not cross the finish line. We finished third by acclamation. The committee boat gave us a toot, signifying our time it was recorded. The race was over.

No need to press, the harbor was a welcome piece of flat water. The sales were doused, and we fired up the engine.

It took hours to go through the lock, the bridge and down the river. After anchoring, we were welcomed to the party going on the lawn. I was offered a beer from a cooler. People also had charcoal grills, covered dishes, etc. They were most generous. It was a lively crowd, mostly young families.

But, as I opened our cooler, I saw the lunch my wife gave me this morning. And the Ziploc bag had a sandwich, a banana, and an energy bar. Suddenly, I wasn't hungry or thirsty – just tired.

I refused the food and drinks and excused myself. Soaked with salt spray and sweat, and I wasn't interested in the awards.

As I entered our front door, my wife asked if I had an enjoyable time. She noticed the untouched lunch. As I sat to take off my deck shoes, I said, "It was a great day," and fell asleep there in the living room.

In Wind in the Willows, River Rat said to Toady "There's nothing better in life than messing around in boats."

CHANGES IN THE WORKFORCE

I was a schoolboy during World War II. But, caught up in the emergency, myself and my brothers worked in my father's steel fabrication shop. Fabricating small parts for Watervliet Arsenal and other war industries,

Retirees, part-time students, and moonlighters turned out to help. The machines we operated required handmade fixtures to be used.

After the war, major changes took place, not only in our shop, but throughout industry the world over. For instance, the railroads and airlines began a great transformation.

The heaviest and latest designs of steam locomotives worked 24/7. Hauling goods and passengers around the clock, in the middle of the night, pounding up the hills surrounding our town. It was said the end of the war stopped the night trains and the birth rate went down.

Before the end of hostilities, plans were in place to replace the locos with diesel electric ones. The new motive power units changed jobs for railroaders. The firemen had a less backbreaking job.

Our hometown mayor even thought of being a fireman on a diesel. The diesel unit could be coupled in pairs and triple operated by a single crew in the lead engine.

Radio and electronic control sped up traffic. Cabooses disappeared being replaced by a FRED device. (Flashing-rear-end device)

Passenger service competed with airlines and lost. Air travel doubled and doubled again in the postwar years. As the aircraft changed, aircrews also changed. Eliminated were navigators, radio men, and flight engineers.

Larger and larger air transport flow long distances, above the weather. Measured in 'passenger-seat miles, fewer pilots were needed.

Manufacturing was modernized. Whereas, during the war, each machine needed an operator. 12 drill presses required 12 workers. Each lathe, punch-press, etc. was run by an employee.

Modern computer-controlled equipment requires few machines to keep the robot system running.

I worked on the railroad, as a trackman. The gang consisted of 20 Gandy Dancers. The pay was a dollar an hour 48 hours – $48.00 Saturday was not overtime. Today the tracks are maintained by a few men operating specialized heavy equipment. Ribbon track in thousand-foot length has eliminated at that clickety clack of rails. Concrete sleepers (ties) last forever.

These workplace changes and many others have specially eliminated low-skill and common laborer positions. Pick and shovel jobs are gone. Visit a job site and you see perhaps one shovel, which is seldom used.

SPRING IS HERE!

Because the winters are long and the summer so short, spring is eagerly awaited by people in the north country. Two noteworthy events herald the start of warm weather. The first is the noise of children playing outside. Because the windows are now opened to air the houses.

We lived across the street from City Hall. This was an antebellum brick, two-story structure with a mansard roof and arched windows. In this single building was a police station, jail, fire house, city offices and courtroom.

The Townhall sat 100 feet back from the street and the concrete paved area in front, was the only open space good for skating. As city workers installed the green awnings with scalloped edges, above the windows and doors – laughing children and the noise of the skating filled the year. The scene was straight out of Norman Rockwell.

The skates being used were the metal wheeled type which strapped on your shoes. Purchased for a modest price at the nearby hardware store. Of course, the happiness was occasionally interrupted by a fallen skater who scraped knee. The injury usually was minor, but very painful.

One mindful occurrence included the butcher's son. The schoolboy begged, pleaded and cried. He wanted a pair of skates. Finally, seeking peace, out of desperation, Dominick the butcher got money for the skates.

Oh, how happy Junior giggled as he strapped on his new skates. Never having skated before, the boy falls. Now crying out loud, he was dragged home by his father. His father shouted, "Don't you put on the skates again until you know how to use them!

The other ritual welcoming Spring was grass fires. You see, the land bordering the railroad properties, which ran the full length of town were not built upon. These open spaces alongside the tracks allowed grass to grow fast and strong. The dried grass presented a hazard to the town and the railyard. So, until after World War II, a spark from the many steam locomotives could cause fires in buildings and the wooden boxcars and sheds.

To alleviate this potential, controlled burns were set, one neighborhood at a time. I should mention the fire department was all volunteer, except for an engine driver, who slept in the engine house.

When grass fires were started, men and boys stood by to contain the fast-burning grass blazes. Each fighter was equipped with a large, heavy straw broom, especially suited for this beating out the fires. Eventually, all the threat was gone.

Soon the seasonal rains would promote green grasses which looked better and were safer.

Cooperation between the town and the railroads was friendly for their mutual benefit. The mayor and city officials and, by and large, all the railroad men lived in town.

One curious thing I learned, concerning the fire department – the present city was chartered by incorporating three small hamlets after the Civil War. Fire companies were private businesses. They provided fire protection to subscribers and insurance holders.

That is the reason that each fire house had a name not a number. The four fire companies were 'Joseph Rings, 'Mink Claxton, 'Hill Hook and 'Latter and Foss.

The earliest alarm system consisted of a five-foot steel ring, usually the tire of a locomotive which was struck with a steel bar.

The rhythm of life was tied to the seasons. The grass fires gave notice that garden plots needed to be spaded and planted. Storm windows were stowed away, and windows were washed, along with a myriad of other chores.

We became anxious for school to end and summer to begin.

RADIO VERSUS TELEVISION

Radio began as telegraphy. Sending messages by Morse code, quickly became a voice, speaking out of the other.

Networks were introduced offering news and entertainment. Practically overnight, housewives, children, and men folk heard radio broadcasts especially to their liking. Housewives could go about housework; men could listen to sports while washing the car. After World War II, FM radio brought concert hall music to our homes.

Television, an outgrowth of radar, was introduced after World War II. Suddenly all the houses sprouted antennas. Viewers sat and watched. But radio had a magic not captured by television – the listeners imagination.

As a boy, I would run at the opening strains of the William Tell Overture. With the thunder of hoofs, "The Lone Ranger rides again!" So, using this magic, follow me.

I went down to the studio where the Lone Ranger Show was being produced. I told the manager I wanted to pet Silver. "What?" he replied.

"I brought a carrot for the Lone Rangers horse."

"Aw gee kid," He replied. "We don't have a horse in the studio."

He said, "we use sound effects."

I was introduced to the sound effects man. He produced horse hoof beats, gun shots, squeaky floors, everything.

His favorite was Fibber Magee's Closet. Used every few weeks, Molly would say don't open the door! Whereupon cascade of pots and pans, broken bottles everything noisy fell out of a garbage can.

The audience roared. He'd do it again. The tinkling was a small bell – that's the directors signal to quit. Life at 78 West Vista brought many laughs

A longtime radio show was Amos and Andy, owners of the Sunshine Taxi. They were played by actors Freeman Gosden and Charles Carrell, two Jewish boys from New York. (They didn't make it to TV. Like Bill Gordon, the voice of Matt Dillon on Gunsmoke. He was a short fellow, built like a fire plug. He didn't make it to TV either.) James Arness was hired as the Marshal of Dodge.

The Hit Parade moved to TV. Snooky Lanson replacing Frank Sinatra.

Television in its infancy had problems but, as time went by, they were resolved. The major improvement was the Cinemascope recorder. Shows were edited and presented in moviehouse quality.

One show that held the largest numbers for the longest time was Edgar Bergen and Charlie McCarthy. Curiously, you could not hire those two characters. If you visited the studio, you'd find a bald Danish man talking to himself. He populated the airways with several voices, often getting in arguments with himself.

But the character I remember most was Mortimer Snerd.

Now, Mortimer was what we might call challenged. He saved his brain by seldom using it. An exchange might go like this.

>Bergen: (calls out) "Mortimer."
>Mortimer: – nothing.
>Bergen: (louder) "Mortimer!"
>Mortimer: – still nothing.
>Bergen: (in a loud vice) "Hey you!"
>Mortimer: "Who? Me?"
>Bergen: "Don't you even know your own name?"
>Mortimer: (Starting to think slowly) "Let me see…I heard it just the other day…"
>Bergen: (After waiting) "You're exasperating!"
>Mortimer: "No. Its shorter than that."
>Bergen: "Oh. How could you be so stupid?"
>Mortimer: "I had a lot of help."

Mortimer Snerd

POLICE RADIOS

Growing up in East Albany, a railroad town in northern New York, was a pleasant and quiet existence. In a town of eight thousand people, the police force was stable and rather uneventful. The officers were neighbors and friends who are knew citizens on a first name basis. They joked about playing pinochle etc. actually was a compliment to the tranquility of the town. As a suburb to the rather large capital city across the Hudson River.

No panhandlers, street walkers or vagrants were tolerated. Miscreants were shooed to the bigger city where a 'tenderloin' district could be found. The river wards, towards the south, accommodated visiting crews from ships in the port. Gambling and vice were to be found and contained in one area.

Before World War II my hometown had only one police car, with no radio. The entire city government was contained in a two-story building. Police and fire house on the ground floor. Mayor, city clerk of court on the second floor.

The two-cell jail was used to store civil defense gear. If necessary, prisoners were kept at the county jail.

There were no women police. Two women volunteers were called in, as matrons, on occasion.

With volunteer fireman, no school buses and a part-time mayor and council, the small-town environment was agreeable.

After the war, modernization began in many ways. The city acquired two new police cars.

Civil defense radios were given to local police and a unique arrangement was made for their use. Rather than build a tower and obtain a transmitter, a system was devised.

The nearby capital city had a tall, powerful transmitter. Used to communicate to nearly 60 patrol cars in the much larger police department. A rented telephone line connected the small police force via the powerful radio facility. To avoid confusion, the two patrol cars were addressed as car 201 and car 202.

My Uncle Mike purchased a receiver which he delighted in listening to police radio traffic, especially at night.

Amongst the big city police traffic of fights, fires, disturbances and calls for backup, there occasionally would be a radio call for car 201. (Transmitter XMTR.)

"Car 201, what is your QRL?" (location)
"Car 201 I'm at J&P."
"I'd like you to code 15."
"Car 201 Roger that, Wilco. Is that with ketchup and mayo?"

TATTING AND SUCH

As a boy, I remember the woman folk staying busy while seated, in conversation or listening to the radio. Their hands were always busy. Handicrafts such as sewing, darning, knitting supplies were at their elbows. The skill of tatting fascinated me.

Tatting required only a shuttle and thread. The shuttle was a tool, made of metal or plastic about three inches long. It was pointed at both ends and carried several yards of thread in its center. Delicate threads of white or colored were used.

I guess patience and dedication are the necessary qualities of an accomplished student of tatting. Nimble fingers made a string of classic designs into a length of lace, to be used as trim on clothes, necklines and cuffs.

Such beautiful crafts were highly prized as gifts and proudly displayed. Perhaps the most treasured items were the few baptismal gowns. Long and dainty, these ceremonial robes were especially ornate, with elaborate lace work. Families used them repeatedly – for generations.

A rare occasion I remember, concerning tatting, occurred after World War II. My father brought home a heavy roll of discarded engineering drawings. These were hand inked linen drawings, known as "F" size, 34 x 44 inches. My grandmother bleached washed and ironed each sheet, making a fine yard or more of broadcloth.

Cut into squares, with narrow hems sown, tatted lace edges were added, to make dress handkerchiefs.

Displayed, not used as hankies, some were personalized with monograms. Receiving a gift of them was always remembered.

The tatting shuttles disappeared around World War II along with the crochet hooks. During the war, knitting sweaters and watch caps of wool for semen became urgent.

Television was introduced. The little screen, demanding attention, robbed us of these handy works and the art of conversation.

A JOKE

Yo ho ho!

A ship of the royal navy visited Port Canaveral one day and held an open house.

My son, David Hill, and I joined a group to tour the ship, following a crewmember as our guide. While in the middle of the vessel, our guy asked us to step aside to let other groups pass.

Sir Roger and his party passed, from bow to stern while Sir Henry and his party passed headed towards the bow.

Our guide remarked, "That's rare, you seldom see two knights passing in the ship."

CLOUDS

Clouds are visible moisture in the atmosphere. They have been studied by scientists and artist for many centuries. My study concerning cloud has been necessitated by my training and experiences as a seaman and an aviator.

As you know, the earth is 3/4 covered by oceans but, 100% covered by the ocean of air.

Air, like water, if left undisturbed with will lay flat. But when stirred up can be violent and destructive. Hurricanes, tornadoes, and blizzards are reported in the news every year.

Floods are caused by weather. Navigators of the atmosphere, as well as sailors, are constantly evaluating their surroundings.

The study of clouds took on great important during World War II. A special base for study was located in Osceola County, Florida. The effects of flying in turbulent air and lightning was studied. This location was selected for several reasons. Central Florida is the lightning capital of the United States. Also, the warmer air supports more moisture. Clouds grow to greater heights and are more turbulent.

Cumulonimbus thunderheads are of special interest. The billowy, vertically developed, giants reach above 30,999 feet.

In a single cloud head vertical air movement, drops of water separated by feet are carried aloft to freezing altitude only to drop down, recoat with water. Increased in size, it repeats the process, traveling up to 300 miles an hour.

Hailstones the size of golf balls are not uncommon – capable of damage to aircraft and property. Across the flatlands of Florida, rows of these giant Cumulus cells stand in a line, along a weather front.

Military aircraft are built stronger than civilian Plains. They are expected to operate under difficult conditions, and they fly from unimproved fields, in bad weather. So, the investigation of all-weather, night-instrument flying was studied.

It has been reckoned that one thunderhead contains more energy than either of the atomic bombs dropped on Japan.

Once, as a teacher, I challenged my class with this problem. Estimate the amount of water being dropped from a single storm head, traveling 15 miles an hour covering 200 acres. Raining four inches an hour.

How many inches acres of water is dropped? How many horsepower for pumps to remove that water, and at the same time? It's a stupendous amount.

Modern long-range aircraft fly at altitudes above the weather. However, they do have to land. Making an approach to the destinations, airliners are directed by weather surveillance radar and onboard systems.

The nose of the airliner is a fiberglass shape housing a radar antenna. Smaller aircraft have a storm scope. This passive system does not send out a radar transmission but listens for the atmospheric interference in towering heads of thunderstorms. Flight crews steer to avoid a rough air and lightning activity.

I worked at Kennedy Space Center, helping the development of the metro data acquisition system. Several improvements to weather science were developed and are in use today.

Lightning doesn't strike twice in the same place, is bunk. We have records of several hundred strikes on one sweep of the scope.

So, leave those beautiful towering Argosies of the skies to poets, painters, and Romantics.

CANDY CANE LODGE

I had worked with a French-Canadian project manager, named Roland. He and his wife had a long career living out of their suitcases, all over the world.

Their last project was a ski resort in Vermont. It presented an opportunity they grasped.

On the long road leading out of town, up to the skiing development, they purchased a two-acre lot. Having become friends with the locals they had no trouble with permits, inspections, and local contractors. Within a year, they open a unique bed and breakfast.

The building had a concrete basement built into the slope. On top was a framed accommodation of four bedrooms, for overnight guests. The owners lived in a well-furnished apartment to the rear, accessed by a private drive which led to the garage and covered porch – their sunroom.

The basement of some 3,000 square feet was a public area. Visitors, mostly young people enjoyed the noisy game room. Ping pong, darts, pool, and a big fireplace swarmed with ski pants and knit sweaters. Some snacks were available. But the owners did not have a license for a tavern.

There was an old-style beverage cooler filled with chilled water, sodas, and beer in tall bottles. People served themselves and no prices were posted or demanded. Only a jar for gratuities was present to cover expenses.

A local welding shop fashioned a huge candy cane of steel tubing, curved at the top. Painted the recognized red and white spirals. It became a landmark.

Roland and his wife loved people and the music and noisy crowds. Guests who did not enjoy the holiday atmosphere were warned, or joined in.

Among the local helpers was a handsome young ski instructor, who was also a charter pilot with his own plane. Being an excellent mechanic, he was well taken care of during the off-season, as the owners spruced things up for the next skiing season.

The pilot, ski-instructor went to South America during our summer months and was a guest at ski resorts in the Andes. He would return to the Green Mountains as the first flurries of snow began.

Another interesting subject is the maple syrup and sugar people always carry home from Vermont. The fact is that New York produces the greater amount of maple syrup. It is shipped eastward across Lake Champlain.

People from New York City go to Vermont to get maple syrup.

THE CHICKEN BUSINESS

 I would say, I backed into the chicken business. It happened in a hasty and unexpected way. The circumstances of time and place caused an impulsive action on my part. It was 1945, the war just ended. I was 13-years old and helping in Dad's welding shop. The phone rang and it was my uncle John.

 Dad's brother, Uncle John, lived outside of town and had run a chicken farm all during World War II. He had decided to change to racing turkeys. My brother Bill and I helped him make the wire runs and other changes necessary for turkeys. Uncle John had changed his contracts with the suppliers and buyers.

 The hatchery that had supplied his chicks guaranteed deliveries. They owed him 100 chicks and had shipped this final number. They arrived by Railway Express. The freight house was a wooden building near the yard office. The freight master called and told my uncle to pick up his chicks now.

 He kept the coal stove going to heat his place. Although it was January, he wouldn't keep a fire overnight for just a few chickens!

 So, my Uncle John called my dad and offered the day-old chicks to anyone who would go get them.

My father, holding the receiver in one hand, shouted into the noisy shop, "Does anybody want a box of chicks?"

I said, "Yes," and brought the little peepers home under a blanket, through the snowy twilight. They were put in a larger cardboard box with water and mash, and a lightbulb for warmth. My mother, sister and younger brothers were curious and excited.

As the subzero winter changed into spring, the new chicks molted and took up much more space. We needed to build a chicken coop and fenced-in yard.

Having little money, we relied on our own resources for materials. Family and friends donated some lumber, hardware another items. But the greater amount of lumber for floor, walls and roof was gleaned from grain doors.

Grain doors, rough lumber like dunnage used across the doorway some box cars, when shipping grain. Stacks of these doors were available as we lived close to the port area and the grain elevators.

We sawed and hammered and made a fine shed of about 12 square feet. A sloped roof eight-feet high on the south side and a six-foot wall at the north. Fence posts, made from car stakes were stuck in the ground eight feet apart. A six-foot-high wire fence kept chickens in and varmints out.

Our flock was no longer chicks, having grown into a few dozen poult. Mortality took its toll. We now fed starter mash and separated hens from cocks.

At 14 weeks we began having eggs and fryers. I oversaw all phases of the chicken business, including killing and dressing. Having a lot of eggs and chicken for the kitchen was great! There were ten people living in our house at that time.

Now the feed bill had also grown. We ranged the chickens on grass and in our large garden. I sold eggs and purchased feed from Grange League of Farmers, a mutual co-op, which sent rebate checks at Christmas time.

Feed sacks of cotton, printed with polka dots or plaid or paisley designs, were sown into aprons, curtains, etc.

Another source for feed, now that the flock was full grown, was the 'empties near the grain elevators. The train loads of grain which came from Canada and the great plains were unloaded and cleaned, sorted, and stored in the great grain silos, to be loaded on ocean freighters for overseas shipment.

The giant machine that would pick up a boxcar, filled with 50 tons of grain, tip it on its side and then raise each end up dumping the cargo into the grating, to be picked up by bucket conveyor. 100 cars would be unloaded, enough to fill a liberty ship in one day.

The empty box cars were put on sidings, awaiting a return west. Fortunately, for me, there remained some few shovels full of grain and dust and in car.

I would split the seams of a gunny sack with my Barlow knife and lay the square of burlap on the ground. Then with a practice toss into the air a shovel full of husks and grain was tossed into the air. A breeze blew the chaff away and a handful of green berries would fall on the burlap. Perhaps I'd reap a few quarts or more of grain to take home.

Once back at the chicken coop, I dump cracked corn, barley, wheat, or rye berries, it didn't matter which, into two galvanized garbage cans. I'd stick my hand deep into the mixture and stir it, before securing the tight-fitting lid.

The coop was cleaned out and disinfected periodically to avoid diseases and pests. The manure was kept to rot and used on our garden.

The seasons marched and nature took its course. The fryers were gone by summers end and now we used hens for stewing. Once a hen lays about 140 eggs it was slaughtered. I still remember lying to my sister. At the dinner table, she wanted to know, "Is this one of ours?"

The years rolled by, and I worked on the docks, on the railroad, and went to sea. When I returned the railroads, the ports, and the barges and ships were no longer there.

Nobody raises chickens anymore. I understand its illegal.

THE TIME OF MY LIFE

I had no watch when growing up. For several reasons. We had little money during the great depression. Secondly, during the war years, production of civilian good was limited.

But my life was well regulated timewise. Getting to school on time and meeting classes was signaled by a bell, rung from the principal's office. Noon was identified by a steam whistle on the round house. This short blast was followed by the carillon in the tower of City Hall, playing the Angelus. Curfew was another signal – a test of the fire alarm system – the clanging of the bell in the fire house nearby, followed by blast on the airhorn atop the hose tower.

Of course, they were those fine wristwatches in the jewelry stores. Names like Bulova, Longines, Wittnauer, and Hamilton; all gold plated and presented in special boxes. I marveled at these mechanisms with gears so small and precise that only the Swiss could build.

To this day, I don't understand the 21-jewel movement. The gold pocket watches, used to keep the railroad running on time, were of particular important in a railroad town like ours. The expression, 'Getting a gold watch,' was associated, with momentous locations – usually retirement.

After World War II, the technologies turned to consumer goods and included new kinds of time pieces. Stainless steel cases. Self-winding watches with luminous styles became common. Timex mass produced accurate, low-priced watches, widely advertised on the new televisions.

Remember John Cameron Swayze? These wind-up watches were cheap and once broken were discarded rather than repaired.

No sooner did the population adapted to the marvels of modern mass production then a revolution took place. Electronic watches, keeping accurate time to a second a year, were introduced. These early model electronic watches still had round faces with the second hand now keeping pace with the ageless minute and hour hands.

With the advent of the 'Moon Race, really miniaturized electronics took over. A battery the size of an aspirin powered an even smaller microscopic electronic chip. These tiny devices kept an accurate cadence by a quartz crystal. Coupled with an LED (light emitting diode) display, the wonders of space-age technologies technology introduced the terms 'analog and 'digital into our vocabulary.

Automatic machinery allowed assembly of devices so small a microscope was needed to view them. Once this billion-dollar industry got rolling, accurate time pieces of unheard accuracy were everywhere.

No longer prized as gifts, or kept as family heirlooms, modern timekeeping is electronic, with digital displays.

Used in jetliner instrument panels, hospital equipment and navigation instruments; indeed, all applications requiring a time function have been adopted.

A few years ago, I was in a dining room with my grandson. Looking up to the large round clock on the wall I tested him by asking what time it was.

He hesitated and slowly walked backward. I noticed that his eyes were glancing aside. He wanted to peek at the digital clocks on the kitchen range or microwave.

No longer do we make appointments as quarter after, quarter till or half past the hour. Indeed, plastic wristwatches, available in variety stores for a dollar, are more accurate than the chronometer in the chart room of the ship I once navigated.

So old Elgins are now displayed in museums or on mantels as curios. Grandpas gold railroad watch with chain is in our strong box. Watches fitted with fobs, some of special importance, were, kept when used, in watch pockets sewn in vests, trousers and coats.

But hey, the watch or fob pockets are no longer offered. Times have changed.

THE WRITE INSTRUMENTS

Growing up during the 1930s and 40s, I recall the common lead pencils and fountain pens used by everyone. A classic yellow number-two pencil of cedar wood, with a red rubber eraser on one end was used by everybody, everywhere. Good for schoolwork, clerks doing ciphers and craftsmen measuring things. Held behind one's ear, the pencil was at the ready for jotting down phone numbers and such.

Benjamin Franklin, as an apprentice, made wooden pencils by hand. They were very expensive.

Pen and ink were used for more formal writing. For a century, straight pens, fitted with a metal nib, replaced the quill of colonial times. These pens were of use only when an ink well was handy, for the constant dipping for more ink after each couple of words. This restricted their use to more stationery locations such as banks and offices.

A great improvement came with the fountain pen. This writing instrument contained a bladder inside the handle, holding a good supply of ink, sufficient for writing many pages, before requiring refilling.

The refilling operation required an ink well and was accomplished by activating a lever on the side of the pen. A cap, screwed over the pen's point, protected the nib and prevented soiling one's clothes.

The mechanical pencil was boom two writers. It didn't need a pencil sharpener. Some mechanical pencils were very fancy and guarded by their owners.

At their zenith in popularity, a set of fountain pen and mechanical pencils were made available in designer sets and nested in elaborate cases. Quite expensive; they were highly treasured, presented as gifts for graduations and such.

In the late 40s, a writing instrument was introduced which revolutionized writing. The ballpoint pen arrived with much fanfare, and everyone had to have one. The original offering was costly and took over as presentation items.

These pens did not require refilling, with watery and messy ink. Instead, after a year or more in common use, the entire penpoint and ink cartridge unit was replaced. TV ads showed this operation wearing white gloves.

Because the ink was mineral oil based, it was somewhat waterproof did not run, thereby eliminating bladders. The ballpoint itself was durable, unlike a fountain pen. It was once demonstrated as being driven into a wooden board like a nail and returned to writing. At their introduction, ball point pens were chained to the counters at store and banks.

But automation in manufacturer produced less elaborate versions. Ballpoint pens are now a as cheap as pencils. They are commonly given as advertisements. Free!

The most ordinary pens aren't even searched for if lost.

A few years ago, it was noticed that ink pens used gravity for the flow of ink. This made them useful only in the down position. They didn't work overhead or against a wall. During the days of the race to the moon, an improved ballpoint pen containing nitrogen gas, was offered as an astronaut pen. It would write in any position, in zero gravity, and these sets are now collectors' items.

I earned a living for many years as a draftsman. For several decades, specialized pencil lead, in special holders were used. Inking sets, such as Le Roy and Wrico, were used on engineering drawings.

The drawings that we produced, to be blue printed, where works of art.

The inking sets, templates and lead holders now gather dust. The computer aided drafting (CAD) systems have eliminated pencil and ink drafting. It was proven, that three CAD draftsman can out produce eight manual draftsmen.

I never made that transition to CAD. That's how I became technologically unemployed – retired.

But, that yellow number two pencil lives on, doing crossword puzzles. Ballpoint pens are found everywhere – in the bottoms of drawers and handbags or abandoned in parking lots.

A REUNION
Or Farewell Gathering

I have never been to a school reunion, or family reunion. My brothers and sisters and I are first generation Americans. Our parents were immigrants with eighth grade or less schooling.

Growing up during the depression 1930s and the war time 1940s there was a lot of old country customs influencing our lives. Perhaps the most prominent family gatherings were funerals. It has been said that the old folks enjoyed the doom and gloom associated with these rituals.

The settings for such affairs were the row houses along the railroad yards. Built at the time of great railroad building, they were typically wood framed, two-story structures, on blue stone foundations. Built on 25-foot-wide lots they had common walls therefore, had windows only at front and rear. A common feature was double doors at the front entrance.

The wide opening doors allowed caskets to be carried into the parlor, for viewing. Typically, a wake was held the evening before the Requiem Mass and burial.

The front room, with the corpse on display, was held to a solemn air. But the rest of the house was filled with a mixture of English and in our family Ukrainian conversations.

Neighbors sent covered dishes of food – many were ethnic specialties. That cortège to the church for services and to the cemetery took all the next morning.

The buffet served not only nourishment, but an occasion to renew acquaintances, make new ones and reminisce. A specialty my mother baked was her funeral pie, made with raisins a sort of mincemeat filling, it was baked especially in winter when fresh fruits were not available.

It seemed as though old folks always died in winter. The frozen ground was not dug for an internment but, remains were kept in a crypt until spring. Folks who traveled from afar were accommodated in our house or nearby relatives.

Prosperity came after World War II and brought changes. The century-old houses had the lathe and plaster walls torn out. New plumbing, electric, and insulation were installed. Gone were the ice boxes and cast-iron stoves. We now had central heat – no more coal scuttles or taking out the ashes.

The fronts of the houses were re-sided and new entrances installed with elegant three-foot wide single doors and electric door chimes.

The neighborhood was filled with automobiles and the trolley disappeared from lack of riders. The granite block street was covered with macadam.

Gone are the giant elm trees and chestnut trees, which lined the streets, and the sounds and smells of the steam locomotives working nearby. Roof tops sprouted television antennas.

The European accents have all but gone from our hearing. All that remains are memories and a few faded photos

JEEP

The word jeep is a slang word born in railroading. Railroad locomotives were of a variety of designs in their developing years. By mid-19th century, two classes of engines were common. The passenger trains were pulled by locos with large diameter drive wheels and heavier, slower freight were identified by smaller drive wheels.

After the Civil War, passenger trains included diners, sleepers, baggage cars etc. Road engines became heavier and more powerful. A class known as general purpose locos were widely introduced. Road crews called them jeeps long before the army decided to mechanize.

Common practice had passenger trains running on main lines according to a time-table and referred as 'number trains with priority over freight. Freight pulled into passing sidetracks were referred to as 'a jeep in the grass.

World War One introduced motor vehicles into the armies. At first trucks and ambulances replaced mules and horses. During the armistice, land armies became mechanized. American forces sought and evaluated several light-weight, cheap vehicles intended for general purposes such as scouting, messengers an emergency transport.

The most promising model for selection was a light one from Bantam Motors. However, when the army requested bids for many thousands, Bantam was out of the running.

Willys provided a rugged, redesigned motor vehicle. Its design featured simple flat sheet-steel body panels and simple assemblies capable of being maintained in battle zones. World War II changed the demand for jeeps from thousands to millions.

Ford motors made most of the wartime vehicles designated GP One. A small number of amphibious jeeps were produced. Designated GPA.

Popularity of jeeps provided a civilian postwar demand, especially from outdoorsmen.

Over time, modernized jeeps, under American Motors, made civilian jeeps that were not used by the military.

Chrysler motors copyrighted the name Jeep, spelled with a capital J. There were also cartoon characters in the comic strips called jeeps. But there will be only one World War II Willys jeep!

A modern Jeep.

HEALTH AND HAPPINESS

I have some observations concerning the amazing growth of the health industries during the 20th century. Professional development, scientific advances, and the social changes in the past hundred years, deserve some study. As a layman, I am not privileged to express any assertions other than my own.

Concerning the terms 'doctor' and 'physician;' the words are now interchangeable. In fact, doctor refers to the achieving the highest degree in education, and physician means a license that has been granted to practice medicine. It is possible to be licensed by a state to practice, and not have a PhD from a degree granting university. It is also not unheard of, for a graduate with a doctorate to not be licensed.

Some students complete the requirements for licensing and don't complete the requirements for the highest degree. Some stay in school to teach or follow other pursuits and have no need for licenses.

The proper title for persons with a master's degree is Mr. Not too long ago people were grandfathered into professional ranks.

This explains the curious careers of Doc Holliday and Doctor Watson. In olden days medicine men had to achieve for themselves, without the protection of a professional society.

Organizations have banded together in recent times and now have great power. Much propaganda and political power has been used for the good of the practitioners. One might say the group 'pulled itself up by their bootstraps.

In recent years, the military wanted to satisfy the needs for medical officers. It was proposed to have a teaching hospital for each of the Army, Navy and Air Force branches. Nothing came of this idea.

The present programs which require candidates to complete undergraduate, post graduate and doctoral training before touching a patient, eliminates the concept of apprenticeship. This causes problems of high student loans and reinforces exclusionary practices.

Now to the technical and scientific advances which have benefited society in so many ways. What we now call space-age medicine, is indeed marvelous.

Citizens taxes have been spent in developing tremendous scientific and technical procedures for the benefit of the few. Advances, such as spinoffs from the space programs, have disappeared into corporate and private sectors not into the public domain.

Keeping knowledge from becoming general in use represents millions, to the select ones. Witnessed a fight cancer with a 'check-up and a check. We all, by now, should know the

12 warning signs without having to send a check.

The gadgetry and its army of technicians have been spawned by advances in electronics. No longer will a little black bag and house-calls be common.

In the name of greater profits, huge laboratories, research tools and great hospital complexes now exist. Because we can now identify parts per billion, not million, and microchip electronics, etc., all sorts of tests, procedures, and proliferating specialists, came into view.

The study of modern healthcare must include the insurance plans. The, now common, plans require employers to earn monies, to be turned over to the health business. No longer do doctors, nurses, technicians must be paid, or not paid, by clients.

This arrangement requires armies of bean counters, shuffling papers in an electronic office, sheltering hospitals and doctors' clinics from scorn. No thinking person can believe insurance companies are attempting to lower costs.

Somewhere, in this train of notions, there are the pharmacies. Older citizens can't recall druggists actually mixing potions and lettering labels by hand.

Well, the counting of pills by a knife is not necessary. Automated manufacturing facilities and the distribution systems should have left the scene long ago. As for the happiness, we should pay attention to the repeated admonishment, exercise and diet used as a disclaimer for every snake oil advertisement on television.

Remember the days before television, during the days of rationing, riding buses, trolleys, bicycles, and walking.

I believe this nation was never healthier than 50 years ago. Farm boys, factory workers, and everyone else stayed away from the hospitals except for a real emergency. We didn't have a health team on our payroll.

A RIDE IN A GLIDER AIRCRAFT

My brother Bud worked as an aircraft mechanic at Albany Airport. As a sideline, he would buy and sell aircraft. I had an Airline Transport Rating license, was looking for work and needed to maintain my ratings. He would let me fly his aircraft.

One summer day, I was returning from Glens Falls to Albany, flying brother Bud's Stinson Voyager, at low altitude. Off to my right I saw a glider. He was approaching a field for a landing. I circled and landed on an adjacent runway. There was no other activity at this airfield except for this group and their gliders.

As I found out later this set of huge runways were built in World War II as a fallback for allied bombers. Tied up in legal matters, the area remains vacant and unused. The glider club squatted on the field on summer weekends.

I was met by several members and welcomed. They admired the ship I flew. I asked about their activities.

"Would you like a ride in a glider," I was asked. I would like to ride any kind a flying machine. So, an instructor pilot introduced me to a primary Swietzer glider.

He sat in back and I strapped in the front seat. The instrument panel had only six flight instruments

– no engine gauges. A big red knob was at the center. At rest, that ship tilted on one wing tip. Upon closing the canopy, a ground crew lifted the wing tip and waved to signal or launch.

The launching of this glider consisted of a long steel wire attached to a hook on the nose. A winch, a half mile up wind, pulled off across the grass. The improvised tow machine was a stripped Pontiac auto fitted with a drum, the size of an oil drum.

The instructor and I were both big fellows, but the ship left the ground after a few hundred feet of noisy rolling on the turf. That takeoff startled me, for two reasons. For one, it suddenly became so quiet we spoke in living room voices. Amazing was the nose high climb. We went up at 30° angle.

Hold it at 45 mph, I was told. The airplanes I flew would fall out of the sky, at 35 mph. We pulled the red handle and released the tow line, at about 899 feet above the ground. What a sensation. It was so quiet you could hear the turn buckles move in their tubes. We did a few shallow turns, always descending. We turned into the wind for landing, at a very shallow 3. °

In motored airplanes, a characteristic landing stall is executed. But the gliders and soar planes are flown at a slight nose down attitude, right into the grass. The rumble of the single wheel was braked by the red handle.

Once we stopped, the ship tilted onto its wing tip. We opened the canopy, and I was all grins.

After thanking them, I flew the Stinson back to Albany. I told Bud we were invited to the next meeting of the glider club to be held next Tuesday at Dr. Bundy's house in Alpaus.

Well, Bud's boss, owner operator of fixed base operation at Albany, knew of the Mohawk Soaring Society. It was founded by Steinmetz at the beginning of General Electric. Among its charter members were Ann Morrow Lindbergh and Charles Lindbergh.

We motored up to the upscale Alpaus Estates where G. E. Management and scientists lived. We were greeted and offered a drink and met several people. The instructor pilot (call me Earl) wanted to talk about the Stinson aircraft. The six-cylinder Franklin engine would make it a great tow plane, especially for their soar planes. My brother, the owner, told them he had already sold the plane to an Eastern Airlines Captain.

We were escorted, drink-less, to our auto. They were interested in the plane, not us.

However, I really enjoyed the ride and a glider. I recommend a ride and a glider aircraft to everyone.

THE RISE AND FALL OF USA RAILROADS

Railroads developed rapidly in the mid-19th century. Roads in the industrialized north generally spread like the web from the east towards the west.

They were able to connect and exchange cars because they were of a standardize gauge. The distance between the rails was fixed at 4 feet-8 1/2 inches, a measure founded by Roman roads, years ago.

In the south, rails varied. Usually reaching from coastal ports inland for shorter distances, varying in track size and gauges. They seldom connected. Most required off loading and reloading between railroads.

This proved a great problem in the Civil War. By comparison, the union was able to move troops and material almost anywhere.

The feat of building the transcontinental railroad, during the Great Rebellion is noteworthy. The parallel development of telegraphy allowed communication and traffic management. Towns grew wherever the rails went.

Many roads built their own engines and rolling stock. This made the interchange of parts impossible. It changed of necessity into standardization.

Two major changes were the safety coupling and the improved air-brake system. Steel rail cars replaced wooden ones. Long distance travel prompted sleepers, diners, and club cars. Long, heavy trains included railway post offices, Railway Express, and even private luxury cars.

Steam locomotives ran on water and fuel. Wood was used originally, but coal bituminous soft coal was the fuel of choice.

Water tanks were spaced along the line. An engine took on water at each 'tank town. High travel in later years, included water troughs between the rails. Trains scooped up water no longer stopping at jerk-water towns.

The final days of luxury rail travel, saw competition with airlines. The finest stainless-steel, air-conditioned luxury rail cars provided world-class comfort and service. But, air travel prevailed. The most modern, most powerful locos and rail cars went overseas.

I guess, people prefer to sit cramped in an aircraft with several hundred fellow travelers, looking at the tops of clouds for scenery.

I traveled south, slept in a Pullman and had lamb chops with mint jelly on linen service while looking out a wide window at the rice fields of South Carolina.

WORLD WAR II
The Loon Weapon System

I worked with Dan Carney at Kennedy Space Center for many years. He lost a leg as a boy and did not serve in World War II. He worked at Republic aircraft, makers of the P-47 'Thunderbolt. Towards the end of the war, the allies needed a siege weapon. Unlike precision bombing techniques, area or carpet bombing was envisioned.

The Germans had such a system, the V-1 buzz bomb, also called the doodlebug. It was cheap, built of non-essential materials, by back-alley shops. Made mostly of sheet steel with no electrical systems, it was powered by a simple pulse-jet engine. That guidance system was vacuum powered by a venturi. Only altitude and direction were held. When the craft ran out of fuel, it plunged to earth. The sound of these low-flying craft terrorized Londoners when the engine stopped.

RAF defense fighters shot down several doodlebugs and a few were recovered for evaluation.

Shipped to Ohio, they were reversed-engineered by Republic. Thus, changing from metric dimensions to American SAE in use. Several thousands were rapidly built and evaluated for use in the field. This was the Loon Project.

However, rapid changes took place and the fate the Loon system was halted. The Germans developed the V-2 rocket, which flew a high ballistic curve from the launch pad deep in Germany also, the Buzz bombs were flown from ski jump sites in France.

The super-secret Manhattan Project was near ready for use. Commando raids in Norway slowed the Germans in the race for nuclear bombs. Just two bombs were used to win the war.

The Loons were shredded and scraped.

Don Carney and I came aerospace workers, in Florida. As cold war veterans, we helped weapons development, then moon projects Mercury, Gemini, Apollo. We finally retired along with the shuttle.

POWDER MONKEY

I was a schoolboy during the great depression, the son of a self-employed detonator. My father had a federal license to buy, store, use and sell dynamite. He found jobs with contractors building roads under WPA contracts.

The market roads in upstate New York were improved by 'cut and fill' techniques. Contractors used rock drills to bore two-inch holes into solid rock formations. The drilling caused heat to build up in the rock, so holes were filled with water and left to be blasted.

Now, dynamite and blasting are not as seen in the movies. Dynamite is nitroglycerin soaked in sawdust and wrapped in wax paper.

It is sold in cardboard cartons. Each month, the boxes must be turned over. This prevents the lower ones from being stronger. Over time, handling sticks may cause the glycerin to quicken one's heart. I was one of five sons, handy to do these chores. Powder monkeys usually stay for only a few years.

The dynamite sticks don't look like broom sticks, and the explosions set off in construction are unlike movies. Under safe procedures, the holes are dried out with a stick and caps placed in sticks. A magneto box is connected on the site.

The ensuing blast is a disappointment. No large orange cloud or shower of rocks. The noise is rather a rumble with little earth tremor or dust.

When necessary, blankets of woven steel cable are used to protect damage to nearby buildings. You see, the process was to break up solid rock to be handled by road building machines. My brothers and I played with the equipment and learned skills that lasted a lifetime. Especially safety. Safety is not a blinking light or a hardhat. It's a frame of mind.

The road projects slowed as the world called up for World War II. My father went into the welding business. Our shop was in the heavily industrial region, where the Hudson and Mohawk Rivers meet. However, we still sold and used dynamite to blast log jams, open graves and remove stumps.

My father was active in civil defense during World War II as a detonator. But that's another story.

THE CATERPILLAR PIN

I worked as an engineering designer in Orlando during the activation of the Martin missile plant.

One day I was assigned to go to a meeting with my lead boss Dan Edwards. He showed up wearing a suit and tie with a caterpillar pin on his lapel.

Being a licensed pilot, I immediately recognized this symbol awarded by the Switlik Corporation. They are manufacturers of parachutes and sent certificates and pins to people whose lives are saved by the use of a parachute. So, I asked about it and he told me the story.

During World War II, he was a pilot flying a C-46 in CBI. (China, Burma India.) One of their routine missions, flying these transports, was as follows. To supply fighter bomber groups in the forward areas with high octane aviation fuel outbound and evacuate casualties on return flights. These operations used an unusual arrangement.

A large bladder, made of material like that used in life rafts, placed on the deck of the cargo space, was filled with a few thousand gallons of gasoline. Upon delivery, the empty unit was rolled up and stowed, making room for the return flight.

The C 46 aircraft built by Curtis was heavier, more powerful than the more widely used DC-3 or C-47. It could fly higher over greater distances. The flight crew usually consisted of two pilots and flight engineer/mechanic.

During this flight, halfway between their base and destination, a problem came about.

Aircraft such as these normally smell of gas fumes. Now the fumes were much greater, and the aircraft flight became unstable. A well-designed and properly flown ship will fly straight and level with little attention. This situation demanded attention.

Upon inspection, the chief reported that a leak in fuel bladder has caused a large amount of gas to run into the bottom of the fuselage. The transport behaved like bucking bronco. If you could imagine carrying a cookie tin full of water, trying to walk and not spill it.

Decisions had to be made fast. As the saying goes, flying is hours of boredom, broken by moments of sheer terror.

The sergeant suggested he punch a hole at the lower most point in the belly and drain the fuel. The hazard of the spark was too great.

"Now we are in a hostile territory. Fly back to base? This is a flying bomb, not welcomed."

So, agreement was to go halfway back and bail out, hopefully in friendly territory.

Turning the disabled craft was noisy matter. All pilots executed single needle-width turn which take two minutes for a circle with shallow banking of the wings.

But that with the liquid slashing about a very delicate 180° was made, with a minimum of banking. Being very delicate on the trim, they reached a point judged to be halfway home. They bailed out and landed in a tea plantation.

The natives and their overseers returned the three airmen to base, where they were debriefed and returned to duty.

The aircraft flew well straight and level, directly towards its base and kept going. It landed, wheels up, in a cleared field, with only minor damage. A recovery crew jacked it up swapped props, etc. and ferried it back to base. It was cannibalized for parts and the fuselage used for storage.

TRIMMING GRAIN

I was in high school during the rebuilding of Europe after World War II. Under the Marshall plan many shiploads of grain were sent to the war-torn nations. America changed to peacetime pursuits and there was a shortage of labor. Boys like me could find work as authorities looked the other way.

The largest grain elevators in the world are located at Albany, New York. Trains and barge traffic from the great plains offload for transfer to ocean ships in the Hudson River.

Now, most people use the terms longshoreman and stevedores interchangeably. Properly, longshoreman stay on the dock and stevedores work on the ship. They may be the same men, but they work under different commands. Loading a ship is like playing three-dimensional chess.

The chief mate is responsible for the hull and cargo. He has a cargo plan which few understand. Not only must a ship be trimmed properly, but cargo also cannot be recklessly over-stowed.

Storage factors must be considered. Putting some cargo on the dock to get at items below, is not wanted. A good supercargo can earn a lot of money.

The business of loading grain is specialized. Most freighters have four levels for storage. The lower hull, lower tween deck, upper tween deck and main deck. Most of the ships have five hatches.

The grain is poured into the hulls while being sampled. Grain, as poured, is in a peaked heap, leaving the sides of the cargo space empty – a dangerous condition if it were to shift at sea. So, gangs of a dozen men in each hold enter the space, armed with large, short-handled aluminum scoops, wearing dust masks. They stand knee-deep in the pile, moving it to fill the space without voids.

If a hold is not filled according to plan, a special method is used. A huge bundle of burlap sacks is tossed into the hold. Each man is issued a handful of foot-long cords, which are kept in his belt. The routine is filling a sack half full of grain. One man holds the ties, while his partner scoops. The millers knot used is quickly learned. To finish the job, the sacks are laid on the level grain two layers. Dunnage, rough lumber, may be placed so additional cargo may be carried into space above.

Me and my buddy Arnold lived across the river and walked several miles to stand in a 'shape up. It was an open secret that racketeers ran the docks. We pay ten bucks for the chance to work. We worked all the overtime, 7 to 11 p.m., got paid in cash, never had a union card.

I went to sea on a merchant ship, was drafted, and later found a life in Florida. The St. Lawrence Seaway opened and caused these grain elevators to be abandoned. Not even the pigeons go there anymore.

If you saw the movie On the Waterfront, they portrayed accurately these conditions.

While I was at sea, Senator Kefauver and the Labor Rackets Committee closed down the cozy set-ups.

But, living in a family of ten people, my mother needed the money, and I'd do it again.

The reason the 'big button boys let us fellows trim grain was because it was an explosive atmosphere...

A BORROWED WHEELBARREL

I needed to borrow a wheelbarrow, so I called my buddy Buck. He and I were shop stewards and moonlighted to supplement to our salaries.

His wife answered the phone and said, "Buck is at the cemetery. He should be quitting for lunch soon."

He and I volunteered there as caretakers, righting gravestones, and fixing them, and the windmill pump, etc.

I saw his pickup and approached. He was finishing digging a 12-inch hole, down to the top of a rough box. Having mixed a bag of secrete, he took a shovelful and eased it into the hole.

He took a shiny half-gallon raid can from his trucks cab. In it were the ashes of a woman who died in Ohio. So, they mailed her remains for burial with her husband.

We cleaned up the tools and I took the wheelbarrow home.

"Where have you been?" asked my wife.

ART VS. ILLUSTRATION

I learned a living as an illustrator, mostly on aerospace engineering projects at Cape Canaveral. All my works are the property of NASA and USAF used in tech publications, kept at Cape Canaveral.

Since retiring in 1998 I began being an artist. Most people don't know the difference. Most art students or hobbyists refer to photos or prints doing what they wish when they wish at their own pace.

Tech illustrators are hired to do what they are hired for. Using acceptable methods, standard materials, work must be accepted on time. Remember, designer and engineering projects have no photos to refer to, there is no need to represent existing objects such as standard hardware or trade goods. Many tech works are now pictorial such a schematics, diagrams, and charts.

Drawings are durable so that printed copies can be distributed. Originals are kept in tech libraries for future reference.

When one considers the number of hours and many disciplines required to create these ink-on-mylar drawings, they can cost huge amounts of money. Never to be seen by the general public.

Since retiring, I have become an artist. Using materials in my own manner, at my own pace, doing what I want to do. Some media, such as pencil, charcoal, and pastels are not to be touched, so they are protected by spraying with a fixative.

Oils the classic material are growing into disuse. Acrylics (house paint) is most forgiving and is widely used.

Watercolors are easily mixed into a multitude of colors. As such, you must continually rinse your brushes and you must plan ahead. Once the paper is dyed it can be a problem to change colors or erase.

Oil and acrylics work can be touched when dry and cleaned when necessary.

More delicate works such as pencil, charcoal, pastels, and watercolors should be framed under glass with proper techniques.

Be creative don't copy a photo or reorganizations you've viewed.

Enjoy.

AUNT KITT
The piano player

Born in 1905 Kitt, was a teenager during the 'Tin Pan Alley days. In the years before radio, people entertained themselves at home. Everybody practiced music. Piano, accordion, violin – you name it. If you didn't play you sang.

For those of us with no talents, there was the player piano. Totally lacking in brain-finger coordination, people merely had to peddle to make piano music.

Aunt Kitt and a few girls played the way their teacher, Professor Schumaker, taught them. New songs were brought to them weekly and would be practiced. Then, when selected, the young girls could play for special occasions

Her notes were transcribed into a long piece of paper, the width of a piano roll. This paper was used to locate square pins on a huge, two-story wheel. This operation produced thousands of copies of piano rolls for the public.

In summertime, before air conditioning, all the curtains would be open. Piano playing filled the street, as we had asked our aunt to play on our upright.

One neighbor, passing said, "I didn't know you had a player piano."

I said, "We don't.

He said, "Well it sounds like my player piano!

I proudly remarked, "Your player piano sounds like our Aunt Kitt."

BUILD A BETTER BENCH

This story was hard to transcribe. I thought it might be fun to show you these scans of the story. Enjoy. WS

①

BUILD A BETTER BENCH
Follow these instructions, and build a strong, lightweight bench, that should last many years. The secret, is in the use of a tubular strut (brace) made of EMT tube.

The materials needed are as follows:
1 ea. 8-foot length of 12X1 inch pine or fir
2 " 1X4 - " " "
2 " 16 inch long 3/4 ft tubing
4 " 1/4-20 X 2 1/2" long carriage bolt
4 " 1/4-20 nuts.
(30 to 40) box nails 2 1/2" long
wood glue

Begin by cutting two 15 inch lengths of the 8 foot (1X12) seat. If you don't have a carpenter's square, use a hard covered book to make 90° (square) cuts. Cut by saw the two 1X4's to the length of the seat. Glue said pieces into a channel shape.

Cut off corners of 1X4's to prevent snags on skin or clothes.

②

Make legs front to 15" long, 2×1.

cut curves with jig saw, make sure you have a snug fit into the top

cut two pieces of 1×1 to glue & nail onto legs

drill 1/4" Ø at center of 1×1 as shown

Do Not drill hole in seat, just yet

Fabricate metal brace (strut as follows)

③

Use a vise (if available) to flatten flatten ends of tubing.
& bend to 45° angle

drill 1/4 ∅
holes for
bolt

· Attach braces to leg first.
Next, center braces on seat, and
drill for bolt (making sure legs
are square (90°) to top.) Tighten
nuts on bolt.
Sand all rough (sawn) ends of lumber.
Finish with varnish or paint.

Enjoy

Paul Hill

HALF TRACKS

I would like very much for a movie producer to make three films featuring half- track vehicles.

As a schoolboy, during World War II, I became interested in these rugged vehicles. The first time I saw a machine with crawler type tracks was a picture of a steam powered logging tractor in the Maine woods. It was an attempt to make logging economical. A railroad without rail and cross ties which had to be moved as lumbering proceeded.

The next mention of half-track machines concerned a French mechanic in the employee of the Russian czar. He made several touring cars with endless tracks at the rear. They were excellent in snow with skis at the front.

With wheels at the front, they could be used off-road and even in mud in summer. He went to France and interested the military in these very useful machines.

Autos and trucks quickly became familiar in cities and developed areas. But the half-track could be useful in off road areas and remote regions, where there were no roads.

As a demonstration, to prove their usefulness, a caravan traveled the silk road from France to China. The stunt was well documented by National Geographic magazine. The public and I were thrilled.

Another publicity expedition sent a convoy of half-tracks across the African desert from the Mediterranean to Timbuktu. Many years later, a retrace of this expedition found their tracks in the Sun-baked mud.

Another half-track adventure was an ill prepared trek through Canadas outback. A dandy and his girlfriend, complete with wines and film crew, mired down and quit. Four-wheel drive vehicles proved superior.

STEAM VERSUS GASOLINE ENGINES

Steam locomotives mostly have two cylinders. When compared to an auto engine they are surprisingly smooth in operation. Each cylinder provides two power strokes for each term of the drive wheels.

Whereas the typical four-stroke auto engine has one power stroke for two revolutions of the crank shaft. To achieve equality with steam, would take 32 cylinders of an auto four- stroke motor.

Another consideration, when comparing these power plants is torque – the starting effort. Steam locomotives can admit full boiler pressure to its cylinders on the first turn of the wheels developing maximum pull of the load. Motor vehicles have gearboxes which permit accepting loads when starting and use highest gear ratios as speed increases.

The steam engine has no selection of gears nor transmission. Starting many years ago in development, several arrangements of valve movements were tried – some not accepted. Trial and trial resulted in standardized valve motion.

Witness the long travels of the valve rod, admitting great quantities of high-pressure steam to the cylinders when starting. Huge blasts from the smokestacks are made.

When running at speed along the main line the engineer adjusts the valve mechanism. The shorter travel of the valve rod then allows a smaller quantity of steam, keeping the train moving at an economical speed.

Railroad workers and hobbyists study the development of locomotives, especially the drive rods and valve mechanism. They've been working hard for over a century.

NO WORKING TITLE

We live on a big blue marble. In fact, our earth is relatively as smooth as an apple. I used to demonstrate on the blackboard a few facts, by using a scale of 100 miles, I would describe big areas holding a chalk in each hand. I am 6'3" tall.

The arcs formed are 80 inches in diameter, representing 8,000 miles diameter. Consider, most satellites are in orbit about 90 to 100 miles above the earth. At our scale, that's around equal to the thickness of your screen door! At the same scale the highest mountains and the deepest oceans would be about a quarter of an inch.

To avoid that now cluttered near earth orbit and its debris, the International Space Station is 300 miles above the earth.

Now, back on the surface, we try to measure plots of land for real estate. Early descriptions of land were not precise, and units of measure varied over time. Stone monuments, dating from pre-biblical times allowed boundaries to be relocated after recurrent floods of the Nile River.

George Washington was a surveyor using a metes and bounds system. Distances rode, chains, fathoms, etc. were not standardized. Many deeds, as recorded were as 'property in possession.

In the vast upper open regions, a few feet or yards was not significant. In built up areas, feet, even inches, were later challenged.

Modern surveyors use new methods in describing a plot of land. The Bureau of Weights and Standards makes sure measuring devices are accurate. A gallon of milk or gasoline, a butcher's scale, or a jeweler's balance are the same from Alaska to Florida. We can buy a gallon of milk anywhere with confidence of its quality and volume. Lumber, such as two by fours may come from several mills but are interchangeable when used.

So, if you measure distances with 100-foot tape measures, they will both be accurate and to a fraction of an inch. Therefore, by measuring diagonals and boundaries and repeating the process the opposite direction, most land deeds as recorded are very accurate. A method of triangular measurement is also used. Since the introduction of the transit, level and plum bobs, precision-built instruments allow very accurate measurement of angles.

Using azimuths (directions angles), accurate buildings and roadways are now common.

Space-age science has introduced the ruby-laser system for measuring long distances accurate to a few thousandths of an inch.

One problem when building or plotting persists. That is man's persistent need to build and measure using right, 90°, angles. Indeed, civilization might be recorded by the practice of plum and square.

In assisting gravity tall columns are pre-built vertically, using a weight on a string a plum lob as a spirit level. Our buildings are built shaped as squares, rectangles, or a combination of 90°corners.

Land measurement followed this practice. Governments, large and small use specially prepared highly accurate maps called cadastral maps for legal and tax purposes.

The famous Homestead Act promoted the settlement of our vast plains. a government survey divided tracts to be settled into mile squares, called sections. Each settler could claim a quarter section.

Since a quarter mile equals 40 acres, each claim included 160 acres. Hence the reference to 'North 40 East 40.

Now comes an interesting feature in land measurements. The plotting square – a rectangular area, on the spherical earth. If you placed postcard shaped papers along the equator, and continue for any distance, long and narrow triangular areas will develop. Therein lay many disputes and work for lawyers.

Two Texan brothers took advantage of this little-known phenomenon called, an error of enclosure. Now we are concerned with mineral rights. The brothers traded mineral rights, with the wildcat oil explorers. For pennies an acre. Many 'dry holes were drilled, and only salt water found.

But these two Texans never drilled a dry well. They hit oil and natural gas each time. The secret was in the land offices records. By taking rights for a few dollars on each one, they waited until oil was found on both sides. Their strips may start out at zero feet, but they increased gradually. A few miles beyond the area it was much wider.

Track mounted drilling rigs only eight-feet wide, backed down the long narrow wedge of leased land to drill amid existing wells.

Science, technology, and modern ways have changed many things. Using great circle measuring for points on our globe, spherical trigonometry is used to plot lines and solve areas instantly by computers!

Mineral rights are often held in public trust managed and by various government agencies. Resources such as lumbering, grading, mining and other resources can a source of public revenue and managed for the general good.

TRUE NORTH

Suppose you want to plot a true north south line for a building a fence or a shed or the like. You may point to the north star, Polaris, at night but have difficulty driving stakes in the dark. During the day, you may have a difficulty knowing when the sun is exactly south of your position. Navigators have special techniques, beyond our situation.

A compass won't help as a compass points to magnetic north, somewhere in Canada. We want 'true north where Santa Claus lives. He has southern exposure on all four sides.

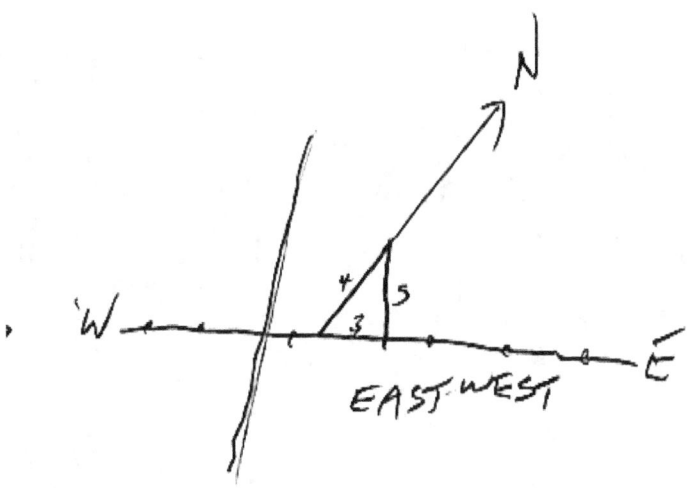

This method may take a little time, but it is accurate enough for most home builders.

On a sunny day, drive a stick or pointed shovel, in the ground slightly tilted to where are you imagine north to be. Beginning in the morning, say nine or ten a.m., place a marker a peg or stone where the tip of the shadow is located every hour or so. By the afternoon you will have a row of markers in a straight east west line.

That was the easy part. We now must construct a perpendicular to this line. One classic method is making a site triangle that is 90.°

It can be made of wood slats or simply measured with a tape measure. The three points of a triangle with sides of three, four and five units, can be seen, yards in meters.

SMALL BOATS THEN AND NOW

On the lakes, rivers, and waterways of upstate New York there were thousands of small watercrafts of assorted designs. But the most common, by far when I was a boy, was the home-built rowboat made of lumber-yard materials.

As a lad, during the Depression, no one had money. Materials were scarce during the war. But, with a few hand tools, no plans, and some determination, a handyman could build a boat like all the others.

Starting out with two 1 x 12 headboards, 12 or 14 feet long. Structural grade number-two was the most common grade in most lumberyards. It usually had a few solid knots and no splits; it was called shelving. Nowadays, what passes for shelving, I would call dunnage – scrap not to be used for construction.

On a table saw, cut the stem from a piece of hard wood, about two feet long. The vee-shape forms the front the boat.

Cut the ends of the sides to a 30° point and fasten with wood screws, spaced about 3 inches apart.

Building the boat upside down. Place a four-foot-long 2 x 4 spreader about halfway to the stern. Bend the two side planks together, using a Spanish windless. Using stout rope, make three or four loops about the ends of the sides. Place a stick at the center of the loops and twist to bring sides together. You can build that transom with framing. Bevel the edges to make close fit. A slight plane to the sides, about 5°, and some work with a block plane.

Now plank the bottom with 1x6 boards across the bottom, spaced the width of a nail to allow caulked seems to swell.

Finish framing with boat upright, seats, risers, inwales, gunwales, breast hooks and knees. Two coats of paint and the last thing - fitting the oar locks. She's ready to launch!

The caulking of seams with wicking and white lead disappeared with use of plywood for building small boats. The ritual sinking of rowboat each spring to swell the seams has not been seen since the 50s.

For some years, an exceptionally good method of producing fine lines and reliable

performance in wood were the molded plywood boats. It required expensive molds and special materials. Not suited for a home-built, it required high-rate production and, being of wood, it required some maintenance for a long life.

With end of World War II aluminum became available. But poor construction and alloys caused aluminum boats to lose favor, except for canoes.

With the introduction of fiberglass, fewer home builders, and the production of lightweight and powerful outboard motors, no one goes rowing in a home-built wooden boat of lumberyard materials.

BALANCE OF POWER

The balance of power is understood was a three-sided triangle representing royalty or civic authority, the military, and the church.

Two sides could join together to correct the improper moves of the third. Say, for instance, if a king became tyrant or mentally incompetent, the church and military would arrange for a regent, and a military coup, would be prevented. Church improprieties could be opposed by the civic authority and the military to maintain the common good for society.

Societies, nations, tribes, congregations, any way to identify a group of people, has a history – a life of itself. Much like a bean, a fish or a club, there is birth, growth, and decline. Some phenomena are long lived, others short. The three-sided balance system may see one side rise in influence for a time, only to be properly correct it and balance restored.

I have lived my life from the early 20th century, into the 21st-century. The world and its people had dramatic changes more rapidly than in times past. From foot paths to wagon trails, railroads, turnpikes, to supersonic aircraft. The mobility for people and goods is amazing.

Industrialization started in the 19th century took hold and really skyrocketed in the 20th century.

Railroads, with the franchises of local and national governments, planned as military networks, provided freight and passenger services everywhere. Natural resources flowed from source to factory to consumer.

Travel on steel rails was paralleled by copper wires and poles for telegraph, and later, telephone. Coal from miners, then oil from wells, led to boom towns. Power plants rapidly lit up cities and powered trolleys.

Farms and ranches became mechanized and fewer people on farms provided more bounty. Mass transportation powered the growth of suburbs. Cities became megapolises.

Private motor vehicles and commercial transport connected all the spaces in between.

A few strong nations took control of many less developed ones. These so-called colonies funded empires and provided raw materials for production and markets for the finished goods. Some nations did not have colonies and felt left out.

World War I sought to put all the recently developed products into the hands of the military. Using machine guns, aircraft, armored vehicles and even poison gas, the war was fought to a standstill to be resolved at a later date. The armistice lasted until 1939.

The roaring 20s saw a wild use of credit. Modernized trading allowed instantaneous trading – old style regulation couldn't prevent a crash. The ensuing 1930s remembered as the great depression, is well documented elsewhere.

The Germans were known to be literate, talented in music and the arts, and an industrialized nation of scientists and technology. Why they listened to Hitler requires special investigation.

Not wanting to get bogged down in trench-warfare mobility was the watchword for everyone. Mobilized infantry and armor, long range artillery, aircraft, and rockets were used by both sides.

So, historians now refer to World War I and World War II as the Great War. Many call them the oil wars.

Observers noticed the Germans needed to use foraging to obtain materials from conquered nations. The high sea fleet was never on the high seas. Germans used merchant raiders and submarines as defensive weapons. Hitler declared war on the U.S. and invaded Russia. A thousand bomber raids on the Romanian oil feels denied him oil. Atrocities brought about war-crimes trials after the war.

The war in the Pacific against Japan was allowed to be fought to deny them sources of oil. Mostly naval battles, with aircraft carriers and island-based bombers proved effective. Japan surrender shortly after two atomic bombs were dropped on the home islands.

The European war ended with Hitler's suicide, and Admiral Karl Dönitz officially surrendering. Japan signed unconditional surrender documents on board the USS Missouri. Collateral damage, that is non-military casualties from disease and starvation as well as non-military property damage, can only be estimated.

NORTH COUNTRY CRAFTS

The Adirondack Mountains are very busy and summers as a favorite area for camping, hiking, boating, etc. During the winter only a few ski developments are very busy.

The vast forests are included in the preserves, dating from New York Governor Teddy Roosevelt, to be forever wild – not developed.

The permanent residents must figure ways to survive through the winter with little or no income. So, each year the many small towns dotting the north country, sponsor 'Mountain Days.

The weekends before and after Labor Day are competing for crowds with street fairs usually as the main attraction. Visitors are offered choices of rodeos, races of all sorts, turkey shoots, you name it. I like the reenactment of revolutionary and Civil War battles. My brothers and I own and shoot black powder guns.

Once the prizes are awarded, the summer folks board up their camps and winter sets in. Only the routes of school buses are plowed. The boat ramps and airports are usually vacant.

Many of the folks spend the summers at various jobs or self-employed keeping an eye on the weather and preparing for the winters.

My Uncle Mike Hill, for instance, was caretaker and handyman on a two-story, lake-front, summer-home for a wealthy doctor. He stayed, alone, in the lakefront house. He lived in the basement keeping a fire and generally doing repairs, as needed.

Over the years he became an excellent boat builder. Each winter he produced one of those more famous guide boats called a Dolly Varden.

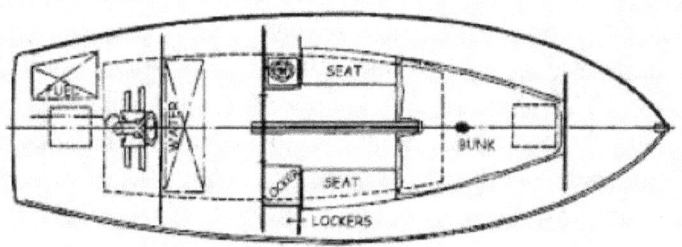

Plan of a Dolly Varden

Steamed rips of green wood became tough and rigid. These pieces were not available in lumberyards. The access to fresh, unseasoned lumber, is the secret of the canoes kayaks and lightweight, strong boats.

Custom built on speculation the finished boats were sought after and brought a good price. If properly kept, some crafts have lasted several generations.

The art of bent wood uses green lumber. It is used by crafters of furniture, hunting bows or snowshoes and sleds. Tied over forms these bent wood pieces were dried, cannot later be straightened.

Flicks are long, thin strips knifed from the board of green wood. Used for baskets and novelties that are decorated to be sold as souvenirs.

Years ago, my brother Bud owned a Ryan Trainer aircraft. It was a two-seat, open cockpit army trainer with a five-cylinder Kinner radial engine. He speculated in used light planes and had a buyer for this one.

So, the day after New Year's, clear but cold, we flew from Albany to Northville located at the north end of Great Sacandaga Lake. At 100 mph, shivering, I could look in the compass crystal and see the snot freezing on my cheek.

Upon landing, the field looked deserted, not a soul in sight. We walked towards the shops built along the side of the hanger. Entering the unlocked door, we yelled, "Hello! Anybody here?"

Two gunshots rang out! Bang and bang again!

"Hey! Stop shooting," in our loudest voices.

"Oh. Sorry." someone answered from the rear of the shop. A man appeared, wearing earmuff sound protectors, and holding a pistol in his hand.

It turned out, the airfield was operated, mostly in summer, by a highly skilled mechanic. He supplemented his income by gun smithing.

As a boy, he bought kit guns and assembled them. Gaining skills and equipment, he began making, bluing, carving, and checkering stocks, and metal etching.

Some highly decorated 'presentation rifles and pairs of pistols with custom-made cases fetch a handsome price. His work is much admired and sought after.

Another cottage industry in maple syrup and products. New York State is the leading producer of maple syrup. Many truckloads are sent to Vermont where people from New York City and Boston go for their maple syrup needs.

Professional model builders turn out museum- quality boats, aircraft, and railroads, including very realistic scenery. Top of dairy and jewelry are also offered the list is endless

Speaking of imagination, the isolation and solitude is conducive to artists of every persuasion. Creative geniuses visit their agents in the big city only occasionally.

You may enjoy a visit to the North country but please don't stay.

THE DAY I MET TARZAN

As a teacher of vocational drafting, I had a responsibility to the students and to the rules and law.

Advanced students, preparing to enter the workforce, can be sent to work in real-life situations. Students can gain a lot working with professionals on the job. But it must be a learning situation not cheap labor for a workplace.

So, one day I received a phone call, from the general manager of the planned Florida Wonderland at Indian River City. He wanted help in preparing flip charts, rough sketches, and such, for a few weeks. I said, I'd be down to check him out after the students left.

The jobsite was being cleaned. Typically, a trailer was set up as a field office. I entered the trailer and introduced myself, he answered all my questions and we made agreements to be followed.

When I was about to leave, he said, "Would you like to meet Johnny Weissmuller?"

I replied, "Tarzan?"

He told me the corporation had hired the movie star, to be their official greeter and host. So, he bade me to follow into the inner office.

There, sitting in a swivel chair, was a heavy-set burger, with that walrus mustache. I must have looked a little shocked. He certainly looked like an elder German, which he was.

We shook hands, and I tried to think of something clever to say.

I told him, "You should know that all the boys in my class lost their singing voices doing that Tarzan yell."

I thought that was witty, and quite original. To that he replied, "I hear that from everybody."

There have been perhaps ten or more actors who have played Tarzan but, in my mind, there is only one real Tarzan in my mind, and in my past.

WEATHER OR NOT

The climate at Americas spaceport is the subject of much discussion. Whether or not to launch depended on information gathering methods that needed improvements if we were to have manned spaceflight.

I'm very proud to have been involved in the several meteorological systems developed in the Apollo program that are much used today. I was a designer/draftsman working on the Metro Data Acquisition Program (MDAP), at Kennedy Space Center (KSC). Engineers were assigned to different systems, and I prepared drawings to be used in the shops and in the field. The solution to the many problems were fascinating. One couldn't help but be proud to be involved.

Atmospheric interference is the static/noise that we hear. Over the years, radio and TV systems have all but eliminated this giving us concert hall listening pleasure.

Well, one of the engineers designed the exact opposite. By filtering out the music and voice, amplifying the AI was made strong. The signal as gathered was not listened to it but presented on a CRT radar screen.

This presentation showed the direction and strength of the ionized atmosphere. When in use, the location and strength of the storms and lightning could be tracked.

Another system concerned lightning. This area is very active for lightning. As the space center is home to a lot of electronic equipment, worth, millions or even billions, it needed protection.

Of course, lightning rods were in place on all structures. But how effective were they?

It was a struggle to properly ground them due to the nature of the soil and we had to use new technologies to make them work.

MY DIFFERENT HISTORY OF MISSILE DEVELOPMENT

I was drafted in 1954 into the Air Force. Among my several buddies, one was an airman from Philadelphia, a graduate of Villanova – he was a patent attorney. He gave me great insight, over time, into inventions and patents. We kept in touch after leaving the Air Force.

Working on various aerospace contracts, having a security clearance as a designer, I assisted engineers, at the Lincoln labs of the office of Naval research – ONR, in Massachusetts. Supporting the design and build of a huge radar, to be used in the Venus and DEW lines, to detect Russian lancets, over the Arctic.

I was sent to Florida in 1957 to work on missiles weapons development. Being vetted for 90 days in September. I told my mother, "I'll be back home by Christmas." I spent the rest of my life in Florida.

Now, let me turn to my understanding of patents and missile development. History has shown that they used flaming arrows, with the range of several hundred yards, in warfare for centuries.

But then, with the invention of gunpowder, rockets came into use. They consisted of a tube stuffed with black powder, much like fireworks. So, the 'rockets' red glare was soon seen.

Ballistic missiles travel a parabolic curve – that means, if I throw a snowball at you can throw one back, following the same curve.

So, I became a 'cold war veteran. The thousands of workers involved took the threats seriously. I worked on the Pershing, Lacross, Bull Pup, Snark another weapon systems.

Let me explain that these systems require the parallel development of a myriad of ground support equipment (GSE).

Now, consider liquid-fueled rocket motors, and their importance.

By explanation, a simple demonstration can be performed. If I take a lead a ball quarter-inch in diameter and throw it at you; you would not be wounded. If I used a sling shot, I will bruise you. But, if I fired 22 caliber bullets from a gun, I could seriously wound or kill. The difference is speed, and a rule of thumb is to double the speed of an airplane or boat requires four times the horsepower. Therefore, the big effort when trying to beat gravity was to get more horsepower without adding weight.

The earliest suggestion of making liquid fueled rocket motors, was by a scientist in Peru. No one took his suggestion to heart. That is, except for a teacher in Massachusetts; Robert Goddard.

Goddard launched several of his hand-crafted rockets, at his aunts farm, setting the woods on fire. Authorities called a halt to his rocket experiments.

Guggenheim, the beer baron of New York City, gave a bunch of money to Goddard to go buy someplace, to launch rockets. Thus, White Sands New Mexico came into play.

Now, addressing the business of patents, and intellectual properties. As science and technology advance, several people may have the same or similar ideas for a new invention. Take for example, the development of airplanes, lightbulbs, automobiles, etc. The pregnant moment may strike several people simultaneously, who then seek to protect their interests.

The patent office is merely a registry. They will not fight legal battles or resolve disputes. Such matters are to be resolved in civil court. Thereby making lawyers rich and inventors poor.

Some suits are legendary, such as Selden versus Ford, or Wright versus Curtiss. Eli Whitney made a fortune using mass-production methods but fought infringement suits over his cotton gin Patent.

My friend, the patent attorney, worked for huge corporations (IBM, Curtiss-Wright). He and a small group of others are permitted to search the registry and make presentations directly to the secretary for approval.

Once a patent is approved, a number issued. It is published in the Patent Journal every two weeks. Scanning the actual words describing an invention, large corporations may show an interest in new products. They have resources (lawyers) to pursue their interests – which could take years.

Back to the missile business. The military showed little interest in Goddards experiments. He had many devices, systems, and formulas which he patented, hoping to gain interest and fortune from his many efforts. Outside the United States, many fractions do not recognize patents and copyrights. Witness present-day concerns over abuses by China.

I suggest Von Braun used the disclosures to further Germany's rocket programs, during World War II. At the end of World War II, Russia and the US used these German experts to catch up during the era called the Cold War.

Early in the industrial age, individual scientists, and helpers, were recognized by name, for their advancements. But Edison changed all of that. The concept of young Tom Edison 'working alone, is fiction.

The wizard of Menlo Park conceived the idea of research labs and the waiver of payment as significant.

Engineers and designers cannot work on research and development projects unless they relinquish their rights for patents. (Or else they don't get a job). When are unique idea is submitted a stipend may be awarded, if it is profitable.

In today's world, scientific and technologies are not named for individuals. Teams work in research to develop the modern marvels we enjoyed.

UNSUNG HEROES

World War II found propagandists found many heroes among the millions of troops in service. The Army, Navy and Marines celebrated their exploits in print, radio, and film.

The uniformed heroes were lionized and awarded medals and citations to be displayed in recognition of their exceptional deeds.

Heroic efforts by thousands of civilian seamen and their ships, totally involved in the war, were not much publicized.

Kings Point Merchant Marine academy, 1952

During the 1930s, the merchant ships were idled and these crews, laid off. However, the ship owners expected to be involved in the coming conflict and they therefore signed long-term contracts with the seamen's unions at 1938 wages. This resulted in circumstances not widely reported or emphasized.

Through 1939, until 1942, ships were being sunk and seaman lost or injured. Sailing without naval or Coast Guard escorts, merchant ships were easy prey to German raiders and U-boats.

Originally, by the rules of war, German warships would stop and search ships allowing crews to take to their boats. Ships with no contraband were either sunk or taken with prize crews on board.

An unfortunate incident in the South Atlantic, caused the beginning of unrestricted warfare, ending stop and searches.

The largest oil refinery in the world, was located on Aruba. Tankers leaving Aruba, sailing to New Jersey were ready targets for U-boats.

Germany declared war on the United States one day after the Pearl Harbor raid, thus involving us in a two-front war.

Few people realize that the thousands of liberty ships and tankers were U.S. Army ships manned by civilian crews.

Now, the practice of sailing ships in convoys, protected by naval gunships, was begun. Anti-sub warfare (ASW) was improved, with the introduction of escort aircraft carriers, sonar, radar, and by arming merchant ships with guns manned by U.S. Marines.

The rapid expansion of the. military, along with the shipbuilding programs, caused a need for thousands of qualified semen – especially officers.

The U.S. Merchant Marine Academy, at Kings Point, New York was expanded with schools in San Mateo, California and Pass Christian, Mississippi. It wasn't easy to find 20,000 plus people who knew how do use a sextant.

Rumors could not be quieted about merchant crews making great money. Especially on troop ships. Army men may have spent 10 days sailing to the front once, but the ship and crew would sail back to the states and do it again. Most sailors are not rich. Most retirees are at sailors homes in Snug Harbor and in paupers graves.

Young cadet midshipmen are in a program of study, which includes one year preparing for sea, then a year at sea on a merchant ship, followed by two years to complete college and attain a maritime certification.

There is a monument erected on the waterfront of the Academy. It is a large plinth surmounted by a full-scale liberty bell.

The bell strikes every half hour in the manner of eight bells for each watch at sea. A bronze placard is mounted on the base. It recognizes the 512 cadets lost during World War II.

THE WRIGHT BROTHERS

The Wright brothers did not invent the airplane. They built and flew the first successful aircraft. They approached the development of their machine and took logical steps solving difficulties others were having.

By studying the efforts of the several other aviation experimenters, they used creative methods to solve the riddles of flight. Even if, in fact, they did not achieve the first successful flights, the many significant contributions would have given them a place in history as pioneers in aviation.

Among their inventions was the wind tunnel. To evaluate the forces of lift and drag they built and used a fan and egg crate structure for testing various forms. Today, wind tunnels everywhere use a Wright Balance to test and evaluate the lift and drag of air forms.

Most pioneers built propellers with flat blades. The rights introduced the concept of a rotary wing. That is, a blade sharpened as an airfoil generating lift, improving thrust.

The beginning of the 20th century introduced the gasoline engine as the prime power plant. The brothers sought a 12-horsepower engine that weighed about as much as a man.

None were available. So, they built the first aluminum-block engine with their own unique carburetion system.

Also, because the boys operated from a bicycle shop in Ohio, they understood light-weight high-strength structures. Tubular compression straps and wired tension members were used. Lift struts, landing wires, drag wires and struts are now universally parts of aircraft.

Both brothers were experienced glider pilots, having made many flights, adding up to hours, giving them knowledge of control of flight.

They built a catapult putting their machines on a rail pulled by a cable for takeoff when the heavy weight was dropped.

The earliest experiment, attempted to steer their airplanes by rudder only – like a boat. The concept of coordinated flight required control of right left up down and tilt to right and left.

Aircraft everywhere are now controlled by rudder pedals connected to a rudder for right and left. A stick is used to move the elevator for up down and, sideways, for roll, the moving the wings down/up.

The roll requirements were accomplished, in the original Wright Flyers by warping the wings. That was later replaced by ailerons.

Like many pioneers, that desire to be first creates secrecy. In aviation in and other fields of endeavor there is a pregnant moment for the creation of a new great advances.

Aviation came about by the lightweight power plants and strong materials for light-weight structures.

It is marvelous that millions of ordinary people can now afford to fly above the weather at near sonic speed.

FEATHERING PROPELLERS
AND PADDLE WHEELS

Small paddle wheels grew larger as more powerful engines were introduced. Large paddle boxes obscured view of pilots, so a bridge was fitted across the boat and hence the term bridge.

Paddle wheels pushed down on the water in front and lifted water at the rear. Screw-type props proved more efficient and eventually forced paddle wheelers into ta different mode.

In the last attempt to make paddleboats competitive a mechanism which feathered the paddles called floats. Each blade remained vertical as it traveled in its circular path. This eliminated the pushing down lifting of the water.

The boats having feathering wheels required smaller diameter paddle wheels and paddle boxes. They proved expensive to build and maintain. So, if you the boats were fit with them. Screws became the common arrangement

During the 20th century there was an amazing development and growth in navigation. The propeller is used to transmit the power of the engines into thrust. As the engine became more powerful and more efficient propellers for improved.

The earliest air screws were made with flat blades. The Wright brothers developed a rotary wing that is an airfoil shaped propeller much more efficient.

A series of designs included changing props for cruise and stunt uses. Then came the ground adjustable prop followed by the airborne prop adjustable in-flight.

Finally, the automatic prop, both hydraulicly and electrically operated, was introduced. Known as the constant speed propeller. It rotated at an RPM set by the pilot.

The most sophisticated design had a feature called 'feathering' was introduced. It was used especially on multi-engine aircraft. When an engine malfunctions, it is shut off. (Think fire, loss of power, or other failure.) The isolated power plant then windmills, that is rotating and causing drag, perhaps causing further damage.

An emergency action will cause the blades of the propeller to turn and be feathered. When effectively managed the blade will be no longer rotate, reducing drag and preventing further damage.

Some aircraft may use reverse thrust upon landing to slow the airplane. The blades on the props are positioned to blow the thrust forward.

We've come a long way from the wooden sticks of World War I. Dead-stick landings are unheard of today. Aircraft designs are much improved for efficiency and safety.

~ ~ ~

ABOUT THE AUTHOR

Paul Hill lives in Merritt Island, Florida. With his wife Shirley. They have been married for 63 years.

They have three children, David, Georgia, and Margaret, and three grandchildren, Anthony, Harper, and Adler.

Paul is an optimistic soul and continues to inspire others by repeating one of his favorite sayings, "Don't sweat the small stuff. It's only small stuff," and, "This too shall pass!"

After reading his stories, you can see why…

A life well lived.